T0220386

Analytics for Retail

A Step-by-Step Guide
to the Statistics Behind
a Successful Retail Business

Rhoda Okunev

Apress®

Analytics for Retail: A Step-by-Step Guide to the Statistics Behind a Successful Retail Business

Rhoda Okunev
Tamarac, FL, USA

ISBN-13 (pbk): 978-1-4842-7829-1   ISBN-13 (electronic): 978-1-4842-7830-7
https://doi.org/10.1007/978-1-4842-7830-7

Copyright © 2022 by Rhoda Okunev

This work is subject to copyright. All rights are reserved by the Publisher, whether the whole or part of the material is concerned, specifically the rights of translation, reprinting, reuse of illustrations, recitation, broadcasting, reproduction on microfilms or in any other physical way, and transmission or information storage and retrieval, electronic adaptation, computer software, or by similar or dissimilar methodology now known or hereafter developed.

Trademarked names, logos, and images may appear in this book. Rather than use a trademark symbol with every occurrence of a trademarked name, logo, or image we use the names, logos, and images only in an editorial fashion and to the benefit of the trademark owner, with no intention of infringement of the trademark.

The use in this publication of trade names, trademarks, service marks, and similar terms, even if they are not identified as such, is not to be taken as an expression of opinion as to whether or not they are subject to proprietary rights.

While the advice and information in this book are believed to be true and accurate at the date of publication, neither the authors nor the editors nor the publisher can accept any legal responsibility for any errors or omissions that may be made. The publisher makes no warranty, express or implied, with respect to the material contained herein.

Managing Director, Apress Media LLC: Welmoed Spahr
Acquisitions Editor: Shiva Ramachandran
Development Editor: James Markham
Coordinating Editor: Jessica Vakili
Copyeditor: Kimberly Wimpsett

Distributed to the book trade worldwide by Springer Science+Business Media New York, 1 New York Plaza, New York, NY 100043. Phone 1-800-SPRINGER, fax (201) 348-4505, e-mail orders-ny@springer-sbm.com, or visit www.springeronline.com. Apress Media, LLC is a California LLC and the sole member (owner) is Springer Science + Business Media Finance Inc (SSBM Finance Inc). SSBM Finance Inc is a **Delaware** corporation.

For information on translations, please e-mail booktranslations@springernature.com; for reprint, paperback, or audio rights, please e-mail bookpermissions@springernature.com.

Apress titles may be purchased in bulk for academic, corporate, or promotional use. eBook versions and licenses are also available for most titles. For more information, reference our Print and eBook Bulk Sales web page at www.apress.com/bulk-sales.

Any source code or other supplementary material referenced by the author in this book is available to readers on the Github repository: https://github.com/Apress/-Analytics-for-Retail. For more detailed information, please visit www.apress.com/source-code.

Printed on acid-free paper

Table of Contents

About the Author ix

About the Technical Reviewer xi

Introduction xiii

Chapter 1: The Basics of Statistics..1

Descriptive Statistics 2

Measures of Central Tendency (Mean, Median, Mode)................................5

Measures of Variability (Range, Variance, Standard Deviation)....................6

Example of Standard Deviation and Variance..8

Computational Example 10

Cleaning the Data Using Descriptive Statistics 13

Summary...14

Chapter 2: The Normal Curve..15

A Statistical Introduction...16

An Important Theorem and a Law..16

The Standard Normal Curve and Its Generalizability Factor 18

How the T-Distribution Converges to the Normal Curve.............................20

Summary...21

Chapter 3: Probability and Percentages, and Their Practical
Business Uses..23

What Percentages Tell Us, and Their Uses 24

Hints to Use to Solve Percent Problems..25

TABLE OF CONTENTS

General Business Examples..26

Example 1..26

Example 2..26

Example 3..27

Example 4..27

Example 5..27

Real-Life Probability and Percent Examples: Markup..28

Example 1..29

Example 2..29

Example 3..29

Example 4..30

Example 5..30

Example 6..30

Example 7..30

Example 8..31

Real-Life Percent Examples: Discount..31

Example 1..32

Example 2..32

Example 3..32

Example 4..33

Real-Life Percent Examples: Profit Margin..33

Example 1..33

Example 2..34

Example 3..34

Example 4..35

Summary..35

Chapter 4: Retail Math: Basic, Inventory/Stock, and Growth Metrics 37

Financial Statements at a Glance 39

Retail Math Basic Metrics 40

Inventory/Stock Metrics 47

Growth Metrics...51

Summary..52

Chapter 5: Financial Ratios 53

Financial Ratios at a Glance..53

Liquidity Ratios 54

Debt or Leverage Ratios..58

Profitability Ratios...60

Efficiency Ratios 62

Summary..63

Chapter 6: Using Frequencies and Percentages to Create Stories from Charts...65

Frequencies: How to Use Percentages.......................................66

Simple Charts: Horizontal, Vertical, and Pie 71

    Horizontal and Vertical Bar Charts......................................73

    Pie Charts 75

Summary..76

Chapter 7: Hypothesis Testing and Interpretation of Results 77

Step 1: The Hypothesis, or Reason for the Business Question.....................78

Step 2: Confidence Level..79

Step 3: Mathematical Operations and Statistical Formulas 80

Step 4: Results 81

Step 5: Descriptive Analysis 81

Summary 81

Chapter 8: Pearson Correlation and Using the Excel Linear Trend Equation and Excel Regression Output 83

Pearson Correlation Defined 83

Hypothesis Testing and Descriptive Steps for a Pearson Correlation 85

    Step 1: The Hypothesis, or the Reason for the Business Question 85

    Step 2: Confidence Level 86

    Step 3: Mathematical Operations and Statistical Formula 87

    Step 4: Results 90

    Step 5: Descriptive Analysis Interpretation of Results 92

Three Examples Using Small Datasets 93

    Step 1: Hypotheses Are All the Same 93

    Step 2: Level of Confidence 93

    Step 3: Mathematical Operations and Statistical Formula 94

    Step 4: Results 105

    Step 5: Descriptive Analysis 105

Summary 106

Chapter 9: Independent T-Test 107

Independent T-Test at a Glance 107

Hypothesis Test 108

Step 1: The Hypothesis, or the Reason for the Business Question 109

Step 2: Confidence Level 109

Step 3: Mathematical Operations and Statistical Formula 110

Step 4: Results 113

Step 5: Descriptive Analysis 114

Summary 114

Chapter 10: Putting It All Together: An Email Campaign 115

Test Goal 115

Method..116

Data Constants...117

Type of Shopper Targeting 118

Time of Year and Duration 118

Cost of Dresses...118

Medium Type 118

Steps to Assess the Success of the Email Campaign 119

Statistics Conducted: Results and Explanations 120

Independent T-Test 1: Conversion Rate Between Models and No Models 121

Independent T-Test 2: Revenue Between Models and No Models 122

Independent T-Test 3: Dresses Sold Between Models and No Models 123

Independent T-Test 4: Orders of Dresses Between Models
and No Models...124

Pearson Correlation by Model: Relationship Between Conversion Rate
and Revenue..125

Sell-Through Rate for Model and No Model......................................127

Average Order Value for Model and No Model.............................128

Total Metrics on Key Performance Indicators for Email Campaign 129

Average Click-Through Rate 130

Type of Model...130

Profit per Dress...130

ROI and ROAS...131

Summary and Discussion on Results...132

Thoughts for Further Analyses 132

Summary...133

Chapter 11: Forecasting: Planning for Future Scenarios 135

Regression at a Glance 136

Establishing Data Collection 136

Predictive Analysis Using the Spreadsheet...................................137

Scenario Analysis...139

Campaign Analysis and Prediction...140

Consumer Analysis and Prediction...141

Summary..142

Chapter 12: Epilogue 143

Index...147

About the Author

Rhoda Okunev is an associate in professional studies at Columbia University's School of Applied Analytics department. Rhoda teaches math and statistics at Nova Southeastern University. Rhoda taught at the Fashion Institute of Technology Continuing Education department in the Retail Analytics department where she created her own Applied Analytics course. She also taught math and statistics at the Fashion Institute of Technology in New York.

Rhoda has a master's in mathematics from the Courant Institute at New York University, a master's in Biostatistics from Columbia University, and a master's in Psychology from Yeshiva University. She also has an advanced certification in finance from Fordham University. Rhoda has worked in portfolio management and market risk for more than 10 years at a rating agency, clearinghouse, and bank. Rhoda has also extensive experience in research and statistical programming at Harvard University, Massachusetts General Hospital, Columbia Presbyterian Hospital, Cornell Medical Center, Massachusetts Department of Public Health, and Emblem Health (HIP).

About the Technical Reviewer

Yvahn Martin is an expert digital marketer, brand manager, and e-commerce professional with more than 20 years of business management experience ranging from small businesses to Fortune 100 companies. She completed a master's in business administration focusing on entrepreneurial finance at NYU Stern Langone School of Business and undergraduate studies at Tulane University. Yvahn has also proudly served for more than a decade on the governing board of Urban Bush Women dance company.

Introduction

If you are an entrepreneur, creator, store owner, or retailer, and you know the products you're selling are stunning and breathtaking, and the brand value your store stands for is extraordinary and relevant, this book can help you maximize your impact using retail analytics. Businesses are expensive to run and costly to manage. You have worked hard to develop and market the products you are selling and to maintain your astute business practices. If the uneasiness of not knowing and not understanding the business analysis tools that could propel your company to the next level is preventing you from moving to the next creative stage to further your career and your company's success, then let this book be your step-by-step practical guide.

This book will help you choose the appropriate path for your company to follow to lead your business to the next level. The spreadsheets in Appendixes A–C (covering accounting, e-mail, and forecasting) are designed to trigger the right business questions to ask and show how to use basic tools of fundamental mathematics, retail math, business metrics, and financial ratios. They are specific to Chapters 4, 5, 10, and 11, respectively, and are available for download at `https://github.com/Apress/analytics-for-retail`.

Business-related probability and proportion questions of markup, markdown, and profit margin are included. In addition, the book will give you the understanding for the need-to-know normal curve and its relevance and importance in statistical methods. It is essential to know how to use these techniques and know how to determine statistical significance so that imperative aspects of your company can be brought to life with relevant ideas and charts. This book will guide you as to how to put it all together as well. Lastly, this book will further that understanding with strategies of forecasting techniques so that tomorrow you will also have the fundamental knowledge and acumen to know where to take your company and how to differentiate it from similar businesses.

CHAPTER 1

The Basics of Statistics

Clear and concise business questions need to be formulated prior to developing a hypothesis and conducting statistics for an online promotion or investigation of some sort. The variables that are collected should be organized in such a way as to answer those business questions, and the statistical techniques that will be used should be known beforehand so that the data to answer the business questions are at hand. Data is just a bunch of numbers until the data is cleaned, organized, and arranged. This part is so essential and much overlooked. Without understanding the basic descriptive statistics and how the middle of the data should look, as well as knowing where the endpoints or tails of your data are and how large the error rate is in the data, a "true" analysis cannot be done accurately. Once the data is rigorously scoured and reviewed, it can, at least, start to tell a clear and consistent message to a targeted audience. That story's journey starts to develop with a clear and understandable business question. With the use of the appropriate statistical analysis and its significant results, and with charts to emphasize the main takeaway to get the points across simply and clearly, the essential and pertinent findings that need to be described will come to life. These findings, which are found from a business question, are the essence of the journey that will help a company realize their direction, purpose, and potential.

© Rhoda Okunev 2022
R. Okunev, *Analytics for Retail*, https://doi.org/10.1007/978-1-4842-7830-7_1

Descriptive Statistics

The term *statistics* refers to mathematical procedures that are used to collect, organize, clean, and summarize data. This is important in retail, because between inventory, customer, and sales data, your overall information sets will be large, and the sheer volume of information can be overwhelming. Cleaning and summarizing your data is key to understanding and interpreting the data accurately. Once that is done accurately, all the statistics are used to create your business narrative and help to optimize your ongoing strategy.

The first concept to understand in statistics is how to differentiate between a population and a sample. A *population* is the entire group that is being investigated. Studying an entire population is usually costly, and it is difficult to impossible to recruit an entire population. Therefore, most investigations and surveys use a random and representative sample or a *random sample*, which is a smaller group of the true population. A randomized sample is a group that represents the population that is to be studied.

For example, imagine a company with 100 employees in the research department and 1,000 employees in the company as a whole. The study's purpose is to assess the quantitative skills of the employees in the research department. In this case, the relevant population is the research department (100 employees), not the whole company (1,000 employees). So, a sample of, say, 30 employees or more should all come from the research department. If 30 employees from across the entire company are chosen by the administrator to participate, this group is not representative of the research department because it includes other departments as well. A company hires individuals from various backgrounds with diversified skills, and the quantitative skills for a sales manager or administrative assistant are not expected to be the same as for a researcher.

To create a random sample of the group of interest, the researcher may take 30 employees from the 100 in the research department. But those employees cannot be handpicked, and you cannot take the first 30 employees because that would not be random. Even when humans attempt to choose items "randomly," it is not accurate. Instead, a possible alternative would be to create a randomly generated table from Excel and to create an algorithm as demonstrated soon in the book or with a statistician for how this sample could be chosen. This way the sample selected would truly be random.

Retailers often use sample sets when analyzing order data for trends by category, when analyzing CRM populations for various studies or surveys, and when analyzing user behavior or visits to websites. For example, if you were conducting a survey of your CRM database, you could randomly generate a stratified sample based on customers who fit certain segments, such as having completed a purchase or having visited the website within a certain period of time. Generating random samples gives you a representative estimate of what is happening in the larger population as a whole. Here are some more statistical differences that should be kept in mind between a sample and a population.

A *parameter* is used to describe a population, and statistics are used to describe that same measure of a sample. The normal curve has three basic parameters and corresponding statistics. They are called the *mean,* the *standard deviation, and the variance.* The parameters for the population are called *mu, sigma,* and *sigma* squared. The random sample statistics are called x-bar, s, and s^2, respectively. These measures describe the population and sample and explain the importance of the normal curve.

The retail business utilizes these measures primarily when analyzing sales data such as orders and daily sales. For example, the mean shows up in key performance indicators such as Average Order Value, Average Units per Transaction, and Average Sellthrough. Variance and standard deviation are building blocks for more advanced applied analytics and appear in later techniques in the book.

One way to create a random sample is by using a random sample generator and then developing an algorithm that would use it. Although this is one simple technique to develop a random sample, another way is to work with a statistician to develop a stratified sample, which is much more complicated. There are many different types of ways to generate random samples, and each type of sample will answer a different question and will give you a different random sample and result. Therefore, it is important to be clear in the beginning of a study which type of representative sample the study is extracting from the population. Some sampling techniques are harder to develop than others, so at times a statistician should always be consulted. The data analysis toolkit in Excel has many different types of random digit generators that could be utilized to create random digits, and the normal curve will be created using the normal random digits in the toolkit in Chapter 2.

Even with a random sample that should yield a representative estimate, there will still be a *sampling error*, which is the difference of mu and x-bar, or the mean of the population and the mean of the sample. A population will never be totally captured by sampling so there will always be some difference between a sample and a population. This difference can be minimized by randomizing the data.

Once a sample is randomized, it is time to start looking at the data to make sure that it is "clean." Moreover, this is the time to start using Excel and to start learning how to use its statistical functions in Excel. Basic knowledge of Excel is expected for reading this book because for the most part only the code is given without a lot of explanation. The following are some types of data problems to watch out for and make decisions about. When encountering complicated issues like these, a statistician may need to be consulted.

The first and easiest errors to detect are the *outliers*, or data that just stands out and does not fit with the other data. Outliers, for the most part, skew your data or maybe do not even make sense. This type of problem could arise from inputting a number wrong in the computer or from an error written in the program. A problem also could then arise from having a subject that should not have been included because it did not fit the criteria. In any event, you want to determine what type of problem occurred and, as much as possible, try to understand and then eliminate these problems by correcting the data or possibly even deleting data, adding dummy data (like the number 999), or inserting information that makes sense (like the mean of the numbers) in the data point when necessary. There is a whole field on missing data, so care should be taken to rectify the situation. However, the variable is altered, so it should be done for a logical reason, and it should be consistent for each variable.

Descriptive statistics are used to summarize, organize, and simplify the data and outliers can be observed. Descriptive statistics are mostly measures of central tendency and measures of variability. These are the terms that the book will discuss now. The next section will address these terms for a sample, not a population, because whole populations are rarely analyzed.

Measures of Central Tendency (Mean, Median, Mode)

The most popular and used descriptive statistics are measures of central tendency. The *measures of central tendency* show where data is most centrally located or where the center of the data lies. These measures include the mean, median, and mode. The *mean*, often referred to as the *average*, is the most popular of all statistics. To calculate a mean, sum up all the numbers and then divide by the count of the numbers. The *mode* tells which items in the data are most frequently occurring. The *median* is the middle number of the sample after all the numbers are put in sequential order.

An example for these descriptive statistics is if there is a sample of five numbers: 3,7,7,4,5.

The mean is the sum of the numbers (3 + 7 + 7 + 4 + 5 = 26) divided by the count (5), or 26/5 = 5.2. The mode would be 7 because that is the most frequently used number in that sample: there are two of them, while all the other numbers occur only once. A sample can have no mode, one mode, or many modes. To determine the median, put all the numbers in order from least to greatest: 3, 4, 5, 7, 7. The number five here would be the middle number. Because there is an odd number of elements, therefore there is a true middle. However, for an even number of items in this list, take the mean of the two middle numbers to determine your median. For example, if the following list consists of six numbers such as 3, 4, 5, 6, 7, 7, the median would be (5+6)/2 = 5.5.

The normal curve is used when the mean, median, and mode are all approximately equal. This is because the data, at this point, is symmetric about the mean: half the data is on one side of the curve's mean, and half is on the other. They are mirror images of each other. The median is used mostly when the data is skewed in some way, like housing data. Housing statistics usually references the median income level or unit price of a home because the very highest prices and incomes are extremely high and would misleadingly skew the average. The mode is used when a whole number is essential, for instance, for the number of children in a household. In this case, 2.5 children per household in a family does not make sense. Instead, it could be written that a typical or mode of a family is three children.

Measures of Variability (Range, Variance, Standard Deviation)

Unlike measures of central tendency, which shows where a centrally located area in the data is, measures of variability are sometimes referred to as *noise, error,* or *volatility* of the data. Variability is a quantitative measure of the difference between one variable and another, or the degree to which the variables are spread apart from each other. The most common measures of variability are the range, standard deviation, and variance.

Returning to the sample 3,7,7,4,5, the range is the highest (or maximum) number minus the lowest (or minimum) number. The range would be 7 – 3 = 4.

The variance for a sample is more complicated. First, take the difference of each X from the mean of the sample X (called the x-bar). Then take that distance of each score from the mean score squared and divide it by n, or count minus one. The formula is shown here:

$$S^2 = \frac{\Sigma(X - \text{x-bar})^2}{(N-1)}$$

The standard deviation for a sample is the square root of variance and is represented by S.

Here are the steps to calculate the variance:

1. Calculate the mean (which is explained in the "Measure of Central Tendency" section).

2. Take every X and subtract it from the mean (x-bar).

3. The sum of the numbers in step 2 will always add up to zero. This is because the numbers center around the mean.

4. Because of this, the square of each number from step 2 is used. Therefore, the variance will always be positive.

5. To calculate the *variance*, take the $\Sigma(X - \text{x-bar})^2$ and divide that whole quantity by (n–1). That number is the variance.

6. The square root of this number is the *standard deviation*. The standard deviation is always positive or zero.

The variance and standard deviation will be discussed further when the normal and t-distributions are described.

Example of Standard Deviation and Variance

For the sample X = 3,7,7,4, 5 where the mean was calculated in the earlier "Measure of Central Tendency" section, the preceding steps are used to calculate the following:

X	Mean	X − mean	$(X - mean)^2$
3	5.2	−2.2	4.40
7	5.2	1.8	3.24
7	5.2	1.8	3.24
4	5.2	−1.2	1.44
5	5.2	−0.2	0.40
26	---	0	12.72 SUMS

The variance for a sample is as follows:

$S^2 = \Sigma(X - mean)^2/(n-1) = 12.72/(5-1) = 3.18$

The standard deviation for a sample is as follows:

$S = sqrt(3.18) = 1.783$

Another method to calculate the variance for a sample is as follows:

$S^2 = S*S = 1.783 *1.783 = 3.18$ (rounding errors can occur)

These three variability terms (range, standard deviation, and variance) are referred to as *noise, volatility,* or *error variables,* and often they could point out potential problems with the data when there are outliers. The range is used when you want to know the distance from the highest number and the lowest number. For instance, if a relay race is timed and the participants want to know the difference in time between the fastest runner and the slowest runners, a range would be utilized. When the range appears too big or too small, the individual numbers that fall near the minimum to the maximum should be examined to determine why this is occurring. Is it an outlier or a subject that should not have been included to begin with?

However, the standard deviation and variance (whose formulas were shown previously) are related, as shown earlier. The variance is more theoretical and, for the most part, used in formulas. The standard deviation is descriptive and could show, based on an approximately normal curve (which is the only case where the curve is symmetric and the mean, median, and mode are approximately all the same), how much dispersion is in the data. So, for example, for normally distributed population let's say the standard deviation is 10 and the mean would be 50. The standard deviation would be subtracted from the left and added on the right of the mean. One standard deviation from the mean would be 60 on the positive (right) side and 40 on the negative (left) side of the curve. Two standards from the mean would be 70 on the positive side (right) of the curve and 30 on the negative (left) side of the curve. This book develops this concept in more detail in the next chapter on normal curve. Thus, the standard deviation measures the distance from the center of the normal curve in either direction because it is symmetric about the mean. The variance is used in formulas to analyze the noise or variability. As explained earlier, variance is the square of the standard deviation.

When the standard deviation is calculated, the first step is to calculate the difference between the mean and its associated set of numbers, X. The sum of the differences will always be zero, which is called the *regression to the mean*. The next step is to square the mean and its associated X so that the sum is not zero, and, therefore, the standard deviation or the variance can never be negative.

These concepts of standard deviation and variance will be elaborated on in the next sections with more examples and may be clearer to you at that time.

Computational Example

Now, it is time to start computing in Excel. Figure 1-1 illustrates the first Excel code program in this book for the measures of central tendency and measures of variability. This is the same data used earlier when these measures were discussed. First, the output in Excel will be shown, and then the code will be displayed.

	A	B	C	D
1				
2		**Descriptive Statistics**		
3				
4	*Data*	Subjects	Scores	
5		1	3	
6		2	7	
7		3	7	
8		4	4	
9		5	5	
10				
11				
12	*Analysis*			
13				
14		**Measures of Central Tendency**		
15				
16			Mean	
17			5.20	
18				
19				
20			Median	
21			5.00	
22				
23				
24			Mode	
25			7.00	
26				
27				
28				
29		**Measures of Variability**		
30				
31			Standard Deviation	
32			1.79	
33				
34				
35				
36			Variance	
37			3.20	
38				
39				
40				
41		Max	7.00	
42		Min	3.00	
43	Range	Max-Min	4.00	
44				
45				
46				
47				
48		**Other Excel procedures**		
49				
50		Sum	26.00	
51		Count	5.00	
52				

Figure 1-1. *Code and descriptive statistics using Excel*

	A	B	C	D
1				
2		Descriptive Statistics		
3				
4	Data	Subjects	Scores	
5		1	3	
6		2	7	
7		3	7	
8		4	4	
9		5	5	
10				
11				
12	Analysis			
13				
14		Measures of Central T		
15				
16			Mean	
17			=AVERAGE(C5:C9)	
18				
19				
20			Median	
21			=MEDIAN(C5:C9)	
22				
23				
24			Mode	
25			=MODE(C5:C9)	
26				
27				
28				
29		Measures of Variabilit		
30				
31			Standard Deviation	
32			=STDEV(C5:C9)	
33				
34				
35				
36			Variance	
37			=VAR(C5:C9)	
38				
39				
40				
41		Max	=MAX(C5:C9)	
42		Min	=MIN(C5:C9)	
43	Range	Max-Min	=C41-C42	
44				
45				
46				
47				
48		Other Excel procedure		
49				
50		Sum	=SUM(C5:C9)	
51		Count	=COUNT(C5:C9)	
52				

Figure 1-1. *(continued)*

Cleaning the Data Using Descriptive Statistics

It is important to input the data without any errors and therefore make sure the data is "clean" using some descriptive analysis. Here are some ways to identify if there are any errors:

- When the mean, median, and mode are all the same number, or at least almost the same number, the data can be approximately normal. This is what you want to see. If the data is normally distributed, then standard normal statistics can be used.

- Outliers skew the data and move the mean of the distribution around and thereby may change the results of the statistics. Charts often are used to point out outliers more easily than just looking at the raw numbers. In addition, the minimum data point, maximum data point, and range are used to identify egregious outliers, which may skew that data. Moreover, the researcher needs to identify what is causing this error and where it is coming from—in other words, whether the problem is from an input data error, computer programming mistake, or real-life issue (such as a customer should never have been included in the dataset to begin with).

- The datasets either have to be from a normal distribution or have 30 or more data points to be large enough to explain the results. Use n (or count) to make sure there are enough data points in each group separately and combined, as well as in each separate group that the researcher decides to analyze.

13

- It is best to analyze the groups that have approximately the same number in each. The data is hard to analyze when there are major differences in the number of counts in between each group.

Note   A discussion of small and large enough samples will be discussed in Chapter 6 when discussing the charts. When there is a large discrepancy between the groups being studied, the data could be skewed and affect the data results for the statistical test. When there is a grossly different sample size occurring, a statistician should be consulted.

- Whether the company you are working at has a small or large data set, the same steps need to be used to clean the data.

Summary

As emphasized in the introduction of this chapter, the business questions of the company need to be addressed initially at the start of promotions, and investigations at hand need to be understood clearly. This is important to do so that you have the right variables in the proper form to aid in answering your research question. Descriptive analysis utilizes the basic measures of central tendency and measures of variability to clean, organize, and arrange the data. This step is one of the most tedious but extremely important. Without clean and well-organized data, no statistics can be accurately analyzed or be understood, and no "true" results will be conveyed. Once the data is massaged, the next steps will be to conduct statistical analysis and analytics. If possible, pictures and charts should be used to emphasize the most important findings. These topics will be discussed later in the book.

CHAPTER 2

The Normal Curve

This chapter introduces the most important and widely used distribution in statistics: the normal curve. The normal distribution is bell shaped and is the most basic of all the distributions and the easiest to analyze and understand. The data that is used to answer your business questions should form a normal distribution, which is when you have a random sample of data large enough or even a small sample that initially comes from a normally distributed population, or both. It occurs when all the basic measures of central tendency are equal—the mean, median, and mode. Or, in real life, these measures should be approximately equal.

The normal curve's formula and distribution are the essence of fundamental statistics, because basically in mathematical terms its function is unimodal and usually smooth and continuous. And, the end points or limits of the normal curve limits approach the positive and negative x-axis but never touch it.

The normal curve can easily be transformed into a standard normal curve (using the Z-score) in order to be able to have a standard normal curve and to be able to generalize the results to other similar studies. The standard normal curve is a special case of the normal curve where the mean is 0 and the standard deviation is 1. In theory, as shown in this chapter, the normal curve may appear confusing and elusive, but it is the lifeblood of all basic statistics. Although it may at first be difficult to understand, no statistics book would be complete without this chapter.

© Rhoda Okunev 2022
R. Okunev, *Analytics for Retail*, https://doi.org/10.1007/978-1-4842-7830-7_2

A Statistical Introduction

The normal curve is sometimes referred to as the *bell-shaped curve* or the *Gaussian curve*. Although in statistics other distributions may be used, the normal distribution is the standard and most important to know, the simplest to understand, and the easiest to use. The normal curve is symmetrically shaped where 100 percent of the data falls under the curve. The data under the normal curve, probability or area is symmetric about the mean, and each side of the curve contains 50 percent of the data so that the two sides of the curve are mirror images of each other.

A more technical definition of what was just said is that the normal curve belongs to a family of special distributions that are stable at their limits. The normal curve is the most important and widely used of all the distributions in this special family, which is composed of random variables $X_1, X_2, ..., X_n$ that are independent and identically distributed (iid). Each X is a random variable in real space with non-negative values. Identically distributed (iid) means each random variable X has an equally likely probability of occurring. The normal curve has a limiting population mean (μ) and a finite population variance (σ^2).

An Important Theorem and a Law

Now, to understand the impact of a normal curve, a theorem and a law must be understood: the central limit theorem (CLT) and the law of large numbers (LLN). The law of large numbers has both a strong law and a weak law. This book discusses only the weak law of LLN because it is easier to understand and enough to prove the point that the mean is robust. The main premise of the weak law of LLN is that with random variables $X_1, X_2, ..., X_n$ that are identically and independently distributed (iid) after a large number of trials yields an expected sample mean (x-bar). Each trial will have the same large sample size (n). The sum of each expected value (x-bar) minus the population mean (μ) converges to 0 (in probability) as the sample size (n) goes to infinity. Since the variables are random and independent and identically distributed, the sequence of observations will vary and so will the expected values (x-bars). The limit of these x-bars, in probability, will converge to the population mean (μ) with the more trials that are run as the same size goes to infinity.

This chapter demonstrates an iid example, using a fair coin where each toss is independent and where heads and tails occur equally often. When a fair coin is flipped once only a head or a tail will occur and the probability of tossing a head or a tail will be 100%. When a fair coin is tossed, both heads and tails should occur 50 percent of the time, or both are equally likely, in probability, and if the coin is flipped with a sample size (n) of 10,000 each time (a large number of tosses), the probability of heads will be approximately 50 percent given a large number of tosses, and the probability of tails will be approximately 50 percent given a large number of tosses. (The probability will be 100 percent for the total number of tosses of heads and tails.) The more tosses (in this case there are 10,000 tosses) of the coin, the closer and closer the expected value of each trial will approach the true population mean (μ) of 50 percent heads and 50 percent tails.

The central limit theorem for the normal curve uses iid random variables and is given two parameters: a population mean (μ) and a finite population variance (σ²). The CLT states, in probability, that using large sample trials drawn, most of the time, yields a distribution around the expected value, the mean. This spread around the x-bar converges to a normal distribution.

The Standard Normal Curve and Its Generalizability Factor

Although the normal curve is the best understood distribution in all statistics, the standard normal curve is a special case of the normal curve, which normalizes that data, and the results can be generalized to other business studies that use the same metrics. The standard characteristics of a standard normal distribution are mean $\mu=0$ and variance $\sigma^2=1$ or statistically written as $N(0,1)$. A normal curve can be transformed to a standard normal curve with the following Z-score formula z= ((x-bar)-μ)/σ. A standard normal distribution plus or minus one standard deviation around the mean or mu is around 34% * 2, or approximately 68 percent of the data or area. (The percent is multiplied by two on both sides of the curve because the curve is symmetric.) The area or data under that curve plus or minus two standard deviations around the mean or mu is 47.50% * 2 = 95%.

In sum, the standard normal curve is used to generalize the results from a study so that the entrepreneur can compare the results to other studies. The normal curve is used in many of the basic statistical procedures used, and it is called *parametric statistics*, meaning it uses parameters such as the mean and standard deviation. The data should be tested by a statistician but could be preliminarily eyeballed in a histogram in Excel.

In the data analysis toolkit in Excel there is a random digit generator where you can specify a standard normal generator with a mean of 0 and a standard deviation of 1. This example uses a series of 10,000 standard normal random digits as similarly described in the example of LLN, with population mean (μ) = 0 and a given population standard deviation (σ) = 1.

Figure 2-1 is a histogram of an approximately standard normal curve that has been generated using 10,000 data points and has a mean of

–0.0084 (which is close to 0) and a standard deviation of 0.9961 (which is close to 1). Since the numbers are random each time a data point is generated, it will vary. Histograms in Excel on the Insert tab are useful tools to aid in identifying whether a curve looks approximately normal.

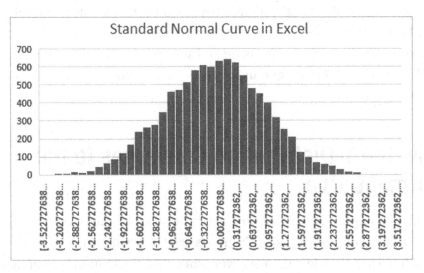

Figure 2-1. *Approximately standard normal distribution generated in Excel*

According to the law of large numbers, as discussed, a sample should be large. So, when a study is conducted, what size sample should be large enough to generalize to a normal distribution? If the original population is normally distributed, even with small sample sizes, the mean will be approximately normal. However, if the population is not normal, most of the time when the researcher draws a large enough random sample from that population, the distribution will be approximately normal. What size sample is large enough? Research scientists have estimated that a sample size of 30 or more is large enough. However, there are more rigorous mathematical and scientific techniques for estimating sample size for a study.

However, what happens if a sample drawn is not approximately normal? What happens if the sample drawn is grossly skewed or does not have a shape and does not approximate to a normal distribution? Then there are nonparametric statistical tests to use instead. For the most part, at least visual comfort with the shape of the approximately normal distribution using a histogram is needed even though this does not seem very scientific. This will be discussed and shown later with diagrams produced in Excel. More rigorous methods for ensuring normality will not be reviewed in this book.

How the T-Distribution Converges to the Normal Curve

When the sample is at least approximately normal or there is a small sample of 30 or more, the t-distribution is used. A small distribution is considered a sample less than 30. Even when this distribution is 30 or more (which is considered large enough to use the normal curve according to many scientists), the independent t-test, which uses the t-distribution, is usually used in statistical packages, including Excel. The t-distribution is used because as a sample gets larger and larger, it approaches the normal distribution. The independent t-test is standard to use because some statistical packages do not even include the z-score test for normal distributions. The independent t-test will be demonstrated later in this book. The z-score test, which was previously shown, is used for population proportions tests or when the population standard deviation is known, which it rarely ever is. It uses population parameters of mean (μ) and standard deviation (σ). The independent t-test is used instead of the sample statistics of mean (x-bar) and standard deviation (s).

When a two-tailed test is conducted with normally distributed data, a z-score of ∓ 1.96 is 95 percent of the area or data under the curve. This shows that the remainder of 5 percent is in the error zone and 2.25 percent is at each end of that error zone curve. This area under the curve is called the rejection region. These ends are called the *tails* of the data under the curve. As a small sample size gets larger and larger for the t-test (which will be discussed later in the book; this statistical test is used for small samples), the distribution will tend to approximate the normal distribution better and better the larger the sample becomes to a certain point. Again, the normal curve is the standard distribution and should be understood before other distributions in statistics can be explained.

Summary

The normal curve is the most basic and easiest distribution to analyze both using calculus and using basic math techniques. This is why no statistics book could be understood without it. It is the one case where the mean, median, and mode are all equal or, in reality, approximately equal. It is symmetric about the mean where half of the data is on one side of the curve and the other half of the data is on the other side of the curve and where the middle of the curve has the most data. The standard normal is a special case of the normal distribution where the mean is 0 and the standard deviation is 1. The standard normal distribution is neat and easy to use and generalize to other studies. Before jumping into any statistical procedure, it is essential to understand that the data must be approximately normal, and although there are some statistical procedures that a statistician can use to measure the normality of the data, even a novice can eyeball the data in a histogram to make sure the data looks approximately right.

CHAPTER 3

Probability and Percentages, and Their Practical Business Uses

This chapter includes basic probability and percent problems that can be used every day in a retail store. The first part of this chapter indicates the rules to follow in probability. Certain rules and fundamental properties are very important to know because they help in understanding the standards of probability. It is also important to know that to change a decimal into a percent, it is multiplied by 100.

The second part of this chapter explains certain types of words that give clues as to how to answer specific retail probability problems. Once the first two parts in this chapter are mastered, it is time to dive into general business examples. These are questions that could be asked on a daily basis in retail, but they are also basic probability questions. After that section, real-life examples are presented on markup, discounts, and profit margins. These sections are shown in a step-by-step approach with the answers to show how to solve these types of questions as they arise in a retail setting, as well as some questions consumers may have.

© Rhoda Okunev 2022
R. Okunev, *Analytics for Retail*, https://doi.org/10.1007/978-1-4842-7830-7_3

Markup is when the store wants to increase the price of a product, a discount is when the store wants to have a sale of some sort and decrease the store price of products, and the profit margin is the amount the store makes from the products. These types of problems are typical and come up regularly when handling merchandise. From the perspective of a retail merchandiser, cost is the COGS and price is the sale prices or MSRP.

Basic math and the order of operations must be understood before going forward in order to understand the procedure for solving percentages. If you feel you need a math refresher, please review Appendix B, which goes over the algebra needed and gives step-by-step answers to the problems.

What Percentages Tell Us, and Their Uses

This section includes examples of percentages and their everyday use in retail. A probability is a percentage of the times an event will occur in the long run under similar circumstances. A percentage can be expressed as a decimal number between 0 and 1 or a probability between 0 percent and 100 percent.

The percent of an event, or P(event) is found as follows: (number of times an event occurs)/(total number of events) = (number of success)/(total number of events) multiplied by 100. A percentage is the number of successes over the total number of possibilities multiplied by 100.

The two fundamental properties of probability are as follows:

- *Property 1*: The probability of an event will always be a number between 0 percent and 100 percent, or the decimal 0 to 1. Zero means an event will never occur, or 0 percent, and 1 means that the event will definitely occur, or 100 percent.

- *Property 2*: The probability of an event or success occurring plus the probability of the event or success does not occur is equal to 1 in decimal and 100 percent in percent.

In other words, the probability calculation will look as follows:

P(Success) + P(not Success) = 1

P(Success) percent + P(not Success) percent = 100 percent

To calculate a probability, in the numerator, the total number of successes needs to be calculated. In the denominator, the total number of possibilities an event occurred needs to be calculated.

For example, say a dart is thrown at a dartboard, and 30 times out of 100 times it hits the dartboard. The probability of the dart hitting the bull's-eye is the probability (30 success/100 total tries)—or 0.30 or 30 percent. This means there is *not* a success of hitting a bull's-eye of 70/100. And, P(Success) + P (not success) = 30% + 70% = 100%.

In retail, it is common for the markup to show a percent that exceeds 100 percent. However, the percentage in mathematics is always between 0 percent and 100 percent. Both the retail technique and the mathematical property are presented in this chapter.

Hints to Use to Solve Percent Problems

Now it is time to look at some more common retail examples. Before that is done, though, Table 3-1 establishes some rules to know for word problems.

Table 3-1. *Word Meanings*

Word	Meaning
Of	Times
Is	Equal
What	X
Decimal to percent	Multiply by 100
Percent to decimal	Divide by 100

General Business Examples

The following are basic probability example questions that may be seen in retail situations every day. The questions are followed by answers and explanations.

Example 1

A retail company has 70 employees, and 20 percent are women. How many female employees are there in the business? How many men are there? What percent of employees are male employees?

Here, 70 * 0.20 = 14 women, and 70 * 0.80 = 56 men. Also, 100% – 20% = 80% of the employees are men.

Example 2

A retail company reports that three-fifth of its employees are men. If the company employs a total of 80 workers, how many are men? How many are women? What is the proportion of men to women?

Here, 80 * 3/5 = 48 men, 80 – 48 = 32 women, and 48/32= 3/2 is the proportion of men to women.

Example 3

768 shirts is 15.3 percent of what total number of garments?

The first thing to do is set up the equation. *Of* means times, and the word *is* means equal to. Therefore, the equation looks as follows: 15.3% * X = 768. Then change the percent to a decimal as in 0.153 * X = 768 and then divide both sides by 0.153. The solved answer is X = 5,020. Remember to round up because a shirt cannot be a fraction.

Example 4

33 percent of the 390 products in a clothing store are shoes. How many shoes are there?

Remember, there are no parts of shoes. Shoes must be a whole number. Since 33% * 390 = 0.33 * 390 = 128.7, there are approximately 129 pairs of shoes. (*Of* means to multiply; *is* means equal to.)

Example 5

A store owner has the following items:

- 12 dresses
- 8 shirts
- 11 pants
- 5 skirts

The total number of items is 12 + 8 + 11 + 5 = 36 total garments, or total possibilities. The following set of questions are some standard probability questions in retail to calculate if one item is selected from many different types of garments:

1. What is the probability the item is a dress?

    Answer: There are 12 dresses (or successes) out of a total of 36 garments (total possibilities), so 12/36 = 33.3%.

2. What is the probability the item is a shirt or pair of pants?

    Answer: There are 11 pants + 8 shirts together (total number of successes) over a total of 36 garments (total number of possibilities), so (11+8)/36 = 52.8%.

3. What is the probability the item is *not* a shirt and *not* a pair of pants?

    Answer: Since probabilities are out of 1, take 100 percent. From the answer in two, subtract 100 percent from 52.8 percent, which yields 47.2 percent.

Real-Life Probability and Percent Examples: Markup

Markup questions are used to increase the price of a commodity from a base or original price. For each percent increase, the formula is as follows: (new price – original price)/original price is used. Here, difference = new price – original price. To change a decimal to a percent, multiply by 100.

Here is the formula for percent increase:

percent increase = (difference/original) × 100

The following are some examples.

Example 1

The average price of a magazine increases $0.85 from $9.75, which makes the resulting price $10.60. What is the percent increase?

Answer: The difference is $0.85, which is the increase. To find the percent increase, the formula for percent increase is used, which is the difference/original price = ($0.85/$9.75) * 100 = 8.72%.

Example 2

The price of membership to a gym increased from $75 to $99. What was the percent increase?

Answer: The percent increase equation is used in this scenario: percent increase = (difference/original) × 100. The difference is $99 – $75 = $24. Then use the equation ($24/$75) × 100 = 0.32 * 100 = 32%. In the end, .32 was multiplied by 100 to get the percent.

Example 3

The price of a book increases 50 percent from $9.50. How much was the increase? And, how much is the price of the book now?

Answer: The equation used here is percent increase, which is (difference/original) × 100. This problem gives you the percent increase, which is 50 percent. The original price is $9.50. Therefore, 50 percent = (difference/$9.50) × 100. The first step is to change the 50 percent to a decimal by dividing both sides by 100. Then the equation will be 0.50 = difference/$9.50. The next step is to multiply both sides by $9.50 to get 0.50*$9.50 = $4.75 = difference or dollar increase. The price of the book after the increase is $9.50 + $4.75 = $14.25.

Example 4

Your rent for your storefront increases by $1,500 per year. The rent is $20,000 per year. What is the percent increase?

Answer: The difference or rent increase is $1,500 per year. The original rent was $20,000. Therefore, when you put it into the equation for percent increase, you have this: (difference price/original price) × 100. In other words, percent increase = ($1,500/$20,000) * 100 = 0.075 × 100 = 7.5%.

Example 5

A clothing retailer used a markup rate of 35 percent. Find the selling price of the dress that costs the retailer (COGS) $45.

Answer: The markup is 35 percent of $45, the cost. Change the percent to a decimal. Remember, *of* means to multiply. So, .35 × $45 = $15.75 is the markup. To get the total selling price, use $45 + $15.75 = $60.75.

Example 6

A retailer pays the wholesaler $60 for a pair of pants. Then the retailer sells the pants in a boutique for $100. What is the markup rate for the retailer?

Answer: The first thing is to calculate the total markup amount. This is $100 – $60 = $40. The same equation can be used here. Rate is percent. So, percent difference = (price difference/original price) × 100 = ($40/$60) × 100 = 0.667 × 100 = 66.7 percent. Therefore, the retailer's markup rate is 66.7 percent.

Example 7

The store manager's income increases by 2.5 percent to $75,000. What is their salary before the increase? This problem needs to be thought through carefully.

Answer: Start with the equation of salary, S, plus the increase in the salary of 2.5 percent, or 0.025. This will yield $75,000. The equations become 1.0 S + S * 0.025 = $75,000. For this problem, the original amount plus the increase leads to the increased salary. 1.025 * S = $75,000 or Salary = $75,000/1.025 or Salary = $73,170.73.

Example 8

A sports store uses a 50 percent markup on cost from the wholesale company they buy from. Find the original cost of sneakers that sell now for $120.

Answer: The final price is $120. Let t be the cost of the sneakers. The equation is the cost of the shoes plus the increase of the shoes equals the total, or 1.0 * t + 0.50 * t = $120. So, $120 =1.5 * t or t = $80 for the original cost of the sneakers.

Real-Life Percent Examples: Discount

Markdown or discounting questions are used to decrease the reduced price of a commodity from a base or original price. In other words, the store loses money when the item goes on sale. For each percent decrease, the following formula is used: (original price – new price)/original price. Here, difference = original price – new price. To change a decimal to a percent, multiply by 100. This is the same formula used for percent increase. The following is the formula for percent discounting:

percent decrease = (difference/original) × 100

The following are some examples.

Example 1

The price of a dress decreased from $110 to $49.99. What was the percent decrease?

Answer: Use the same equation of percent decrease, which is (difference price/original price) × 100. Take the difference between $110 – $49.99 = $60.01, which is the decrease or difference. Then use the equation percent decrease, which is ($60.01/$110) × 100 = 0.5455 × 100 = 54.55% decrease.

Example 2

The payroll for a department store decreased $45,000 from $600,000. What was the percentage decrease?

Answer: Again, the same equation is used: percent decrease = (difference/original) × 100. Therefore, the equation becomes ($45,000/$600,000) *100 = 7.5% decrease in payroll.

Example 3

A retailer sells pants for $1,000. The pants are marked down to $400. What is the discount rate?

Answer: First, find the markdown amount or the difference: $1,000 – $400 = $600.

Here is the equation for percent discount: (difference price/ original price) * 100. This is ($600/$1,000) * 100 = 0.6 * 100 = 60% markdown percent or discount rate.

Example 4

A necklace is not selling well at a store, so the storekeeper discounts the price from $200 to $50. The store bought the necklace for $110. What is the discount percent? What is the percent profit or loss margin?

Answer: Here, solve for the difference between the original price ($200) and the discounted price ($50). Use the equation (original price – discounted price)/original price = percent discount, which is 75 percent in this case.

The difference between the discounted price ($50) and the price the store bought the necklace at ($110) is the profit margin, $110 – $50 = $60. So, $60/$110 * 100 = 0.5454 * 100 = 55 percent loss in profit margin, indicating a loss of $60 on the product sale.

Real-Life Percent Examples: Profit Margin

To determine the profit margin, calculate the difference between the price and the cost. Although retailers do mark up items over 100 percent, the profit margin is always a portion of the ending price. The price cannot go below 0, but the cost of the item remains fixed so profit margin can be positive or negative. Then use the formula ((Sales Price or MSRP) - (COGS)) = Profit Margin. Profit Margin% = Profit Margin/(Sales Price or MSPR) * 100.

The following are some examples.

Example 1

A festive dress is marked down 20 percent after the holidays. The sale price is $60. What is the original price the store sold the dress at MSRP? What is the margin if the original COGS is $35?

Answer: Only the sale price and markdown percent are known. Let X = original price of the item. Since the sale price is 20 percent marked down the item price is 80 percent of the original MSRP, or 100% – 20% = 80%. Therefore, the equation becomes (X * 80%), which is (X * .80) = $60. The original price the store sold the dress for was X = $60/.80 = $75.

Margin is the sale price minus the cost of goods sold (COGS). This is $60 – $35 = $25 profit the store made on the dress. The profit margin percent on this dress would be $25/$60 * 100 = 41.7% profit.

Example 2

A fancy velvet and silk skirt is marked up 80 percent on the cost of goods. The marked-up price is $240. What is the COGS it cost the store to purchase the skirt? What is the profit margin?

Answer: Only the increased price and markup percent are known. Let X = COGS of the item. Since the increased price is 80 percent marked-up, the item price is 100% + 80% = 180%. Therefore, the equation becomes (X * 180%), which is (X * 1.80) = $240. The COGS will be X = $240/1.80 = $133.

Profit margin is the marked-up cost minus the COGS. This is $240 – $133 = $107 profit, which is a 45 percent profit margin.

Example 3

A computer is marked up 200 percent before Christmas. The original cost of goods the store paid is $400. What is the marked-up price? What is the profit margin?

Answer: Here, only the original cost (COGS) the store paid for the computer ($400) is known, and the increase is known (200 percent). Let X = the difference. Since the increased percent is 200 percent, the equation is written 2.0 = X/400, which means X = 800. The marked-up price the computer was sold for is $800.

The profit margin is $800 – $400 = $400, or 50 percent.

Example 4

During a store clearance the price of a shirt is discounted 80 percent. The original price the shirt was sold for (MSRP) was $2,000. What is the profit margin if the COGS is $500?

Answer: The original price, discount percent, and COGS are known; the sale price must be calculated before you can determine profit. The sale price is the MSRP *(1 – 80%) = $2000 × 20% = $400. Then, take $400 – $500 = –$100 or 20% loss on COGS, which will be the profit margin.

Summary

This chapter demonstrated practical problems for solving real-life business issues that may be encountered daily by retailers and consumers. It showed, using a step-by-step approach, in a question-and-answer format, how to approach and solve fundamental principles in probability. Grammatical clues in the words for solving word problems were demonstrated as well. Then the chapter showed real-life scenarios of calculating markups, discounts, and profit margins, which are types of questions that often come up in retail settings.

CHAPTER 4

Retail Math: Basic, Inventory/Stock, and Growth Metrics

This chapter gives a framework for how to understand, calculate, and assess retail math. Examples of a retail company's financial history and statements (in the form of an Excel spreadsheet) can be found in Appendix A at `https://github.com/Apress/analytics-for-retail`. This should be downloaded prior to reading the chapter. Each metric explained in the following sections will refer to specific tabs in the spreadsheet to illustrate the example.

The example company's two-year financial history is given via an income statement, a balance sheet, and a cash flow statement, which are featured throughout the chapter. These are designed to aid in the understanding and analysis of how an actual company has performed in comparison to its previous year, which illustrates a short-term performance trend. The basic metrics section features key performance indicators (KPIs) for the main functions of how a company works, as well as metrics for profit. The inventory or stock metrics section indicates the amount of inventory that is left over and when to stock and restock the inventory. The growth section allows the company to have a good idea of its growth potential.

The way to diagnose the well-being of a company is through analyzing different metrics for different reasons and then summarizing the analysis into

© Rhoda Okunev 2022
R. Okunev, *Analytics for Retail*, https://doi.org/10.1007/978-1-4842-7830-7_4

a conclusion. This chapter will go over the basics of retail math. To calculate retail math, an understanding of basic math, fractions, averages, and percent is essential. (Please see Appendix E for help with basic math if needed.) Retail math includes basic metrics, inventory and stock, and growth metrics of a company performance. *Basic metrics* emphasize sales or revenue, some profit KPIs (key performance indicators), average sales, transactions, and markups of a company. (Chapter 3 covers how to calculate markups); *inventory* means how the merchandise at the company is selling; and growth metric describes how the company is growing from one-time period to another.

Each KPI will be associated with the values on the spreadsheet of the financial statements for a fictitious dress merchandise store. At the end of the chapter, retail math components will be reviewed and analyzed with recommendations for the future of this merchandise company. Chapter 5 will discuss the financial ratios and why and how they are useful. Let an accountant review your calculations because there are many variations for some of the formulas. For instance, some formulas could use quantity or dollars. The metric names can vary as well depending on the variables being calculated. Each accounting spreadsheet will have its own unique nuances depending on many factors and products being sold.

The metrics reviewed in this book are some of the most fundamental and basic metrics you will need to observe yearly, and sometimes on a daily, weekly, and/or monthly basis. Since each season has its own challenges, it is a good idea to record what they are. For instance, August and September have back-to-school and fashion weeks. Starting in September through December, the company is getting ready and displaying items for the holiday season, and January and June are sample sales and annual sales with big price markdowns. In addition, along the way, there are long governmental weekend holidays and cultural holidays to remember as well as competitive events such as Amazon Prime Day.

Check with your vendors to review manufacturing timelines to produce and ship their product to you at the time requested; you cannot meet a sales goal if you do not have the inventory to support it.

Financial Statements at a Glance

Financial statements describe pertinent pieces of information to understand the well-being of a company, which includes the income statement, the balance sheet, and the cash flow of a company. The metrics described over the course of this chapter are what help a company measure its financial health. The income statement shows the company's revenue (referred to as *top line*) and expenses during a certain period, usually once a year. Although you usually have to report once a year to the IRS, performing this analysis once a month gives you a better perspective on your business activities. It shows how to calculate the income and earnings before income, tax, depreciation, amortization (EBITDA), but if a company does not grow or maintain its profit, then the company may be in debt. This statement emphasizes whether the company made a profit or loss during a certain period as indicated.

The balance sheet is a "snapshot in time" of financial balances. It is usually analyzed for a company monthly, quarterly, or six times per year. It refers to the assets, liabilities, and equity ownership of a company for a nonpublic company. The total asset is equal to the total liabilities plus its owner's equity. The assets are financed by the liabilities (borrowed money) or equity (owner's or shareholder's). A company's desire is to have the equity large and the liabilities as small as can reasonably be expected—that is the company's profits).

The cash flow statement illustrates the money that flows in or out of the business during any given measured time period, typically monthly. This statement categorizes the sources of revenue and expense into areas useful for the company such as revenue from investing versus revenue from operations, or debt-related expenses versus operations-related expenses. This can give the observer a clearer perspective on whether funds and resources are being allocated effectively.

Retail Math Basic Metrics

Basic metrics are also known as *key performance indicators* (KPIs) of a company. Basic metrics introduce how the basic functions of a company are working by its gross sales, net sales, average order value, units per transaction, ticket price cost of goods sold (COGS), initial markup and gross profit percent, average unit cost, average unit retail, operating profit, and net profit. These KPIs assess whether a company is achieving its basic goals and performances.

The book will now go over each basic metric and give the formula for the KPI, relevant information about the KPI, why it is important, and the values from the financial statement for the dress merchandise company:

Gross sales = Gross revenue or at times referred to as top-line sales.

Gross sales are relevant to retail business and show how much merchandise the business is selling. Gross sales are the total proceeds of all the sales within a time period. To calculate gross sales, the company needs to sum up all the sales receipts. Gross sales do not include the operating expenses, tax expenses, or other charges. This metric is on the income statement.

The gross sales increased from 2XX0 ($1,130,000) to 2XX1 ($1,408,455), which shows that the company is expanding its top-line form of sales. This is always a good marker with which to start.

Net sales = Gross sales – (Discounts + Returns + Allowances)

Net sales is the revenue earned by a company for selling its products related to the company's pertinent operations and reveals the strength of a company. This metric is sometimes referred to as *revenue* and is used to calculate many other KPIs because a company must bring in revenue in order to turn a profit. This is the retail value of any product after discounts, returns, and allowances for missing or damaged goods are removed. The sales reported on the income statement are net sales. Some of the ways to increase revenue are by distinguishing the business product from others in the market, increasing the customer traffic via advertising, increasing the frequency of transactions and the quantity of product in every transaction via recommendations and cross-selling, and post-purchasing email reminders. You may also increase the average unit sale or average sale, increase the price of the popular merchandise, or even better negotiate with the wholesale companies about instituting better terms for the cost of the goods sold and allowing the company better terms to sell the product at a more competitive price. Anniversary sales, shopping holidays, and other special events provide regular opportunities to showcase your brand value and acquire new customers.

Net sales are increasing, which is a good initial sign, from 2xx0 ($1,072,500) to 2xx1 ($1,336,842). Although this is a first sign that the company is doing well, there are many other factors with the other metrics that will need to be assessed before coming to a conclusion about the well-being of the company.

It is good to graph gross sales against net sales and see if their movement is going in the same direction and both are improving. The owner of the company can then see if the sales after all the discounts, returns, and allowances are still increasing.

For both years, the gross sales and net sales are increasing, which indicates just by eyeing the numbers that sales are increasing and thereby improving.

Average order value (AOV) = (Net sales in dollars) / (Number of transactions)

The average order value measures customer behavior and looks at the average amount of money a customer spends every time a customer places an order. This variable describes the average dollar per transaction in a period of time and helps you evaluate the pricing strategy of your retail product and its long-term value. In other words, once the company has an idea of how much customers will spend on average for a product, the company can develop strategies to increase the number of items sold per transaction and use promotions and bundle the products together in order to increase the number of goods sold. This metric shows the average amount earned from a customer over a period of time, usually monthly, weekly, or daily.

From 2xx0 ($536) to 2xx1 ($637), the average dollar per transaction increased. This is a good sign and shows the company is learning how to increase the volume per sale.

Units per transaction = (Quantity sold) / Number of transactions

This measures the average number of items that customers purchase in any given transaction. The higher the number, the more items customers are purchasing per transaction in a visit. It is usually recommended to calculate this metric daily, weekly, monthly, and yearly, and to see if there is a change from year to year.

The number of units per transaction decreased slightly from 2xx0 (4.26) to 2xx1 (3.98). This is an important metric. It indicates from the AOV that customers are spending more per transaction but with fewer items in their shopping cart. Therefore, the reason may be that customers are buying higher-priced goods but fewer items. It is important to look at this daily to see what is going on, why it is happening, and what you can do about it. Here, the company may want to bundle products in order to reduce the price of the products that are selling together and/or work with the company that is producing the goods to see if the popular products could get a better price for the cost of goods and thereby lower the price of the merchandise.

*Ticket price = Quantity sold * MSRP*

This ticket price is the sum of the tickets of all the products in the store. MSRP stands for manufacturer's suggested retail price or original ticket price, and the average ticket value takes into consideration any markdowns and discounts on items sold. The Average ticket price can be used instead of MSRP.

The ticket price has increased from 2xx0 ($1,193,500) to 2xx1 ($1,461,250), which illustrates an increase in potential profitability based on the growth in initial markup year over year in our example. If the initial markup remains the same and the retail value increases, it indicates a higher quantity of items available for sale. The company needs to make sure that their retail product is essential to the buyer and different from its competitors so that it will sell more products at its maximum possible ticket price.

Cost of goods sold (COGS) = (Beginning inventory valuation at full price + Net purchases + Cost of labor + Material and supplies + Other manufacturing costs like freight and shipping) – End-of-period inventory at full cost

Advertising, fees to website hosts or other technology vendors, agencies, and other sales-related costs often go into a separate line item from COGS but are important to include when calculating EBITDA and cost of sales. COGS and EBITDA are in the Income Statement.

This is an important variable to the retailer because it usually is the largest expense and it includes how much it costs the company to produce the items in the store.

The cost of goods increased from 2xx0 ($184,000) to 2xx1 ($200,000). The object is to keep this number as low as possible. It is a good idea after the company is more comfortable with the distributor that it tries to reduce this cost and negotiate a better price for the supplies and materials, etc.

Initial markup percent = [(Ticket price– Cost of goods sold)/
Ticket price]* 100

The initial markup percent needs to cover the wholesale cost, as well as payroll, taxes, and day-to-day expenses, and the cost of running the business of the company. Chapter 3 discusses calculating markups.

The initial markup percent increased from 2xx0 (85 percent) to 2xx1 (86 percent) in response to the cost of goods also increasing to compensate for the additional cost. This may have slowed the volume of sales (Quantity sold) from year 2xx0 (8,525 units) to year 2xx1 (8,350 units), but resulted ultimately increased profit margin dollars. A goal of the company is to bring the cost of goods down without lowering the quality of the product.

Interesting enough, for both years, the markdown percentage remained constant at 5 percent from 2xx0 to 2xx1. And, with this markdown and discounts included, the amount for 2xx1 ($1,190,827) was better than 2xx0 ($939,077). This amount indicates that perhaps the company under-priced the goods in year 2xx0 and gave the products a better initial price in the second year so that there were fewer markdowns and discounts.

Gross profit = (net sales) - (cost of goods sold)

Gross profit margin = [(Net sales) - (Cost of goods sold)]/ (Net sales)*100

Gross profit margin includes the direct cost of the company's profit and is an important metric to determine the company's profit because the cost of goods—the largest expense—is subtracted from the equation. Direct cost includes labor expenses and material expenses. It does not include the indirect costs such as advertisements, which is a sales and marketing expense, or rent and utilities, which would fall under capital expenditures or operations. The goal is to make the cost of goods less expensive and the gross profit higher. The company wants to see this metric increase.

The gross profit margin is used primarily as a mechanism for managing top-line pricing. The gross profit margin should not fluctuate but at times does due to changes in pricing and retail markdowns. If it fluctuates too greatly, it may be a sign of poor management. It should be relatively stable unless the company is liquidating products at a lower negotiated price, e.g., wholesale liquidation. In this case the gross profit margin would decrease based on negotiated wholesale or dropship arrangements with fixed commissions. For each business sector, the gross margin needs to be determined based on industry standards for your type of company. A good gross profit margin for online stores is thought to be around 46 percent to 65 percent.

Even with the cost of goods sold higher, the gross profit increased from 2xx0 ($888,500) to 2xx1 ($1,136,842) and so did the gross profit margin from 2xx0 (83 percent) to 2xx1 (85 percent). This is a positive position the company is in. It shows that the gross profits of the company are improving.

Average unit cost = (Cost of good sales)/ (Quantity sold)

This is the average amount paid per unit of an item. A company will want to lower the cost of an average unit in the end. They can do this by figuring out the average unit cost and then seeing where they are able to cut costs. Each product has fixed costs that do not change and variable costs that could be adjusted. It is good for the company to always be on the lookout for better ways to adjust the variable costs.

The average unit cost increased from 2xx0 ($22) to 2xx1 ($24). The amount did not increase too much, but the company needs to find a way to decrease this cost per unit by trying to reduce the amount the company has to pay for the cost of goods.

As mentioned, the MSRP is the initial price recommended by the producer or brand to maximize revenue.

Average Ticket price or retail price is the final price that the product sells for to the end customer, including any markdowns or discounts.

It is important to reduce additional costs or expense that are not necessary to get the merchandise into inventory and ready for sale, like shipping and handling fee. Accurate stock prices are important for a company, and moving the product from the stock room into a sale is imperative.

The average ticket price of the goods increased by only $10 from 2xx0 ($135) to 2xx1 ($145). Customers do not want to see the price of their clothing jump too high, which is a way to maintain and increase the number of customers who buy the product.

Average unit retail = (Net sales or revenue in dollars) / (Quantity sold)

This metric is the average selling price of an item and informs the company about the microeconomics of the product: how much a buyer is willing to spend on the product, how many items the customer is willing to buy, and if the cost of the item is too high or too low.

The average unit retail increased from 2xx0 ($126) to 2xx1 ($160). It shows that the customer is willing to buy the higher-priced product.

Operating profit = gross profit – total operating expenses

The operating profit is the income earned from the performance of the core business operations of a company. This means it excludes from the calculations interest and taxes, as well as earnings in which it may be invested in other businesses or other investments. In other words, the direct and indirect costs of the business are included. This metric should be watched closely to see if it is improving, which shows how the company is operationally profiting. If the company needs to borrow more funds to keep it working, this will negatively impact operating profit metric and will show up in this KPI.

The operating profits increased from 2xx0 ($548,500) to 2xx1 ($718,642). This signals that when interest and taxes are excluded from the calculations, the operating expenses are improving and the core business is doing well.

Net profit = net sales – (cost of goods sold + operating expenses + taxes + interest)

Net profit margin = (Net profit) / (Net sales)*100

Net profit is also referred to as *net income* and is the money left over after expenses are paid. This measure tells the company the amount left over at the end of a period of time.

The net profit margin metric, which comes from the income statement, shows how much profit the company makes after the cost of goods sold, taxes, and operating expenses (both fixed and variable cost) are subtracted from the net sales in the numerator. In other words, it is how much of every dollar in sales a company keeps from its earnings. The net profit margin is subject to fluctuation based on both the promotional and discounting activities and seasonal investments in marketing and other operations expenses such as software subscriptions, shipping and warehouse costs, and head count. The net profit margin becomes EBITDA when depreciation and amortization are added into the calculation; EBITDA will be discussed in the next section.

The net profit margin shows the profits a company would have to use if the company had to cover increases for some of the fixed costs of the company, as well as cover the variable costs of the company. A *fixed cost* is a cost that remains stable over the business activity. A rent expense is a good example of a fixed cost because it is usually a set amount at each quarter. And, that amount will not change as the market may change.

A *variable cost* is an expense that changes as the business activity changes. The raw material to manufacture the product may vary each time the company purchases the material.

The goal of a company is to make a net profit month after month. This metric is essential to track because it shows if the company is making enough money for the sales and contains the operating costs of the company. It will also show that the company is stable and on a trajectory with growth in the future. A company wants to see at least this metric maintain itself and increase over time.

A good profit margin is around 20 percent; 10 percent is considered acceptable, and 5 percent is a low percentage. The higher the better. The net profit margin remained stable and was very high from 2xx0 (48 percent) to 2xx1 (47 percent).

Inventory/Stock Metrics

Managing the inventory of the store is no easy task, and the stock control needs to be reviewed often. This section discusses inventory management metrics, which aids in helping management decide how to make informed decisions on the inventory. The metrics included in this section are start of period inventory, units on hand, stock dollar EOM, fill rate, stock/sales, week unit sold, weeks of supply, beginning week on hand, and sell through. Inventory turnover is in the financial ratio chapter. It is essential for a company to know what is happening with the inventory and how the stock is handled and sales are produced.

Start of period inventory = This measure tells the retailer the total inventory at the beginning of a period and can be calculated in units, cost, or retail dollars depending on the audience for the report. Units are typically used by retail planners to allocate shipments to retail stores, warehouses, or wholesale customers. Cost is typically used by finance to estimate inventory liability, while retail dollars on hand indicates risk to sales revenue as well as sales potential and is used by the sales and marketing teams. It is generally assessed at the beginning of the week or month.

For the year, this number remained the same from 2xx0 ($5,000) to 2xx1 ($5,000).

End-of-period inventory = This is a measure of the total inventory a retailer has at the end of a period and can be calculated in units, cost, or retail dollars. It is generally assessed at the end of the week or month. Typically retailers would use the start of period or end of period for their reporting, but not both. The difference between the starting and ending inventory for any period is the sell through, whether in units, cost, or retail.

In general, when referring to inventory or stock, it is the value of goods and is referred to in terms of dollars.

Units on hand = Starting inventory – Quantity shipped + Quantity purchased

Units on hand lets the company know how much merchandise the company has until the stock runs out. This KPI lets the company know if they have enough merchandise to cover all the inventory and if the company will have enough merchandise to replenish the inventory once it is sold out.

This is the current units of stock the company has. This metric increased from 2xx0 (1,475) to 2xx1 (2,250). The units on hand are increasing and may mean that the company needs to work better with the supplier to bring lead time down, know their customer so that the company is aware when they will be buying specific products, and place orders with the supplier once an order is placed.

*Stock dollar EOM = Units on hand * Average ticket price*

This is the amount of cash that the inventory is costing the company to hold at the end of the month. This is calculated at the line item or SKU level to accommodate pricing differences between products and then summed to a total.

The stock dollar increased $127,125 from 2xx0 ($199,125) to 2xx1 ($326,250). This is not a positive number; however, if the cost of goods price decreased, perhaps the ticket price could be reduced, and this metric would look better.

Fill rate = (Number of orders shipped) / (Total number of orders or Transactions)

*Fill rate percent = Fill rate * 100*

This is a forensic measure that is useful to help predict future orders. This measurement helps you understand how well the company is able to meet customers' demands. It is the key to improving customer experience and to redefining a wholesale inventory management and fulfillment center. Fill rates should be as close to 100 percent as possible. For fill rate, the company needs to also assess the dead stock, loss and damaged products, and products that are obsolete now. Sometimes orders and transactions are interchanged. Transactions includes returns.

The percent went up slightly from 2xx0 (94 percent) to 2xx1 (97 percent). This indicated that the company seemed able to meet the customer demands for the products.

Stock / Sales = (Stock dollars EOM) / (Net sales)

This measures inventory on hand, which is based on the previous months of sales. This metric shows how much inventory was needed to achieve the sales in a month's time. The lower the number, the more sales will have moved because the denominator is sales. This measure is very important to small and medium-sized companies because inventory is one of their largest expenses. It is important to maintain a good balance between a well-stocked inventory and selling enough merchandise to move your inventory into purchases from storage in order not to lose money. A good stock to sales is in the range of 0.16 to 0.25.

Stock to sales increased from 2xx0 (0.19) to 2xx1 (0.24). Although both are in the range, 2xx1 is getting close to the boundary, and it should be assessed to find out what is going on. Again, if the cost of goods decreased, this number may improve. Also, the company could have more presales and promotions to help cut the stock to sale number.

Weekly units sold = Units shipped/52 weeks in a year

The weekly units sold decreased from 2xx0 (164) to 2xx1 (159).

Weeks of supply = (On-hand inventory units)/ (weekly unit sales)

This measures how many weeks it takes for merchandise to sell and how long the inventory will last given the average current rate of sales. This measure is a forward-looking metric and helps the company forecast to assess what strategy to use to reduce the inventory by advertising this product with promotions or discounts.

Here, the numbers increased as well from 2xx0 (9) to 2xx1 (14) showing that there is more inventory than needed. This shows that the company is carrying too much supply and it should be reduced.

Beginning week on hand = (Starting inventory)/(Weekly units sold)

This metric is the average amount of time it takes a company to sell the inventory it has. For small and midsize companies, this metric is a snapshot of the company's health because inventory is a big expense for the company. Investors may look at this criterion to see the health of the company. The lower the number the quicker the company is able to turn its investment into revenue. Too much or too little to sell should be watched carefully because the company does not want to run out or have too many products.

The beginning week on hand increased from 2xx0 (30) to 2xx1 (32). This means that the inventory is out for approximately a month for both years.

Sell through = (Units sold)/ (Units sold + Beginning of the weeks units on hand)*100

This metric is usually analyzed by retailers weekly and is a very similar metric to turnover. Regularly priced merchandise and reduced merchandise with regular prices are commonly analyzed together to compare sell-through velocity. This is a comparison measure of the amount of inventory a retailer receives from a manufacturer or supplier to what is actually sold. This amount indicates how much of the supplies the company has gone through and if the company needs to restock any of the items. A good sell-through rate is between 40 percent and 80 percent. The higher the percentage, the better.

Sell-through decreased from 2xx0 (85 percent) to 2xx1 (79 percent). The sell-through rate for this company is pretty good, which indicates the company is selling through its merchandise at a good pace.

Growth Metrics

The growth metrics focus on improving the operations of the online sales and payment to increase revenue. This section consists of two metrics: last year total sales and build. This metric aids in making sure that the company is on a positive trajectory.

Last year total sales = Net sales/build

Last year total sales increased from 2xx0 (893,750) to 2xx1 (1,072,545). This is a good sign.

Build or trend = (This month's total sales)/(Previous month's total sales)

This is a measure of the change from one month or week from the previous month or week. Typically, it is ideal if this time keeps increasing. The build increased from 2xx0 (1.20) to 2xx1 (1.25); that is a good sign.

Now that we know the definition of each KPI, what each metric does, and why it is important, the next step is to dive into the analysis of the retail metrics for the dress company. However, it does not show all five years, which is when the company was incepted, of financial statements the company. Only the past two years are shown, and it appears that the flow from the past two years is positive. This is a good sign indicating that the core of the business has enough money to invest in new equipment, profit, and grow. This is a positive sign of potential growth. It also shows that the company has enough money to pay for the previous debt and potential future loan agreements. The cash flow statement predicts the future health of a company's ability to pay back loans. It lets the administration know that the company has enough money to pay its expenses.

What the company needs is for buyers to negotiate a decreased cost of goods with suppliers while expanding the inventory base. This will help enable the company to increase profitability. Once that is accomplished, the company's task is to increase its sale prices and maintain its merchandise costs. This cost optimization process may help increase items per transaction and have a ripple effect with profit.

Summary

This chapter illuminates basic retail math, inventory or stock, and growth metrics with the aid of Appendix A. Through a fictitious retail company's example, this chapter covered how to use retail metrics in a constructive and useful way to analyze and start telling a story of a retail company. The next chapter will demonstrate how to look at the financial ratios using the data from the same real-life company, and then it will explain how to sum up financial ratios information.

CHAPTER 5

Financial Ratios

While retail math is useful for merchandise companies, financial ratios are used for all types of corporations. These metrics are useful to further analyze retail businesses, as well as assess all companies' health and determine where their strengths and weaknesses lie. This chapter examines metrics that assist in inspecting whether the example company has enough cash to cover its expenses in the short term, whether the debt and obligations are too large to cover the expenses, whether the company is turning a profit, and whether the company is efficient in using its resources. The ratios explained in this chapter are illustrated with the financials used for the previous chapter's fictional company on the Cash Flow and Financial Ratios tabs of Appendix A, which has already been downloaded.

Financial Ratios at a Glance

Financial ratios, like retail math metrics, enable a company to further determine its financial health. Financial ratios measure how much liquidity, debt or leverage, profitability, and efficiency a company has. These ratios are important because they help a company assess the cash it has to spend, how quickly it's spending on lines of credit or cash reserves, and whether it has enough cash to cover regular obligations such as rent, debt service, and operations expenses.

© Rhoda Okunev 2022
R. Okunev, *Analytics for Retail*, https://doi.org/10.1007/978-1-4842-7830-7_5

The liquidity is the current or short-term assets of some form over current liabilities of some form. Current means a time period of a year that is considered a short period of time. It specifies whether a company has the money to pay off its current debt and more.

The debt, sometimes referred to as the *leverage ratio*, indicates the risk that a company is carrying. The liabilities vehicle of a company may be debt, loans, interest rate, or taxes. Without a company having enough liquidity to cover the debt and more, they will be in a financial situation where they cannot pay back the credit they may owe.

The profitability ratio shows how much profit the company is able to keep from the amount of income earned. This ratio will help a company know how it grew and how it can plan for the future.

The efficiency ratios are sometimes referred to as *activity ratios*, and they give insight into how well the company manages its operations and sales activities. The goal of this metric is to show how the company produces income through the effective use of the company resources.

For these ratios, it is essential to know when the numerator (the number on top) and the denominator (the number on the bottom of a fraction) are the same and the result is 1. When the numerator is greater than the denominator, the result will be greater than 1, and when the numerator is less than the denominator, the result will be less than 1. For instance, when you look at the current ratio, to make sure the company is able to pay down its current liabilities, the result should be greater than 1. This will show that the assets are at least as large if not larger than the current liabilities.

Liquidity Ratios

Liquidity ratios measure how much time it will take a company to repay an obligation. It is the ability to meet the debt of the company as it is due. There are three main liquidity ratios: current ratio, quick ratio, and cash ratio.

Current ratio = (Current asset)/(Current liability)

This current ratio indicates how strong a company is and how well it is investing its capital. The current ratio is sometimes referred to as the *working capital ratio,* and it helps a company understand how it could cover its current liabilities within one year. Both current assets and current liabilities are on the balance sheet. Current assets include cash, accounts receivable, inventory, and other assets that are expected to be liquidated or turned into cash in less than a year. The current ratio is a liquid asset or a vehicle to move around money quickly. It measures the ability of a company to pay off short-term liabilities quickly or those debts due within one year. The current liabilities include accounts payable, wages, taxes payable, and the current portion of long-term debt.

A company with a current ratio less than 1 may not have the capital on hand to meet short-term obligations that are due. A company with a current ratio greater than 1.5 is, in general, doing well. This number may depend on what sector the company is operating under, how liquid the assets really are, how easily it may be to refinance its debt, and what the plans are to use excess assets. The current ratio, if less than 1, explains at a point in time that a company cannot pay its current debt; however, once those payments are received, it does not mean the company will not be able to pay the debt. Therefore, it may be difficult to compare different companies using only this metric because of what it measures.

The current ratio is increasing from 2xx0 (2.64) to 2xx1 (3.65). Both are good indicators. Both these KPIs for the company have current ratios greater than 1.5 and show that the company has enough short-term assets to cover the short-term debt.

The quick ratio has two formulas that both yield the same results.

Quick ratio = (Cash – Inventories – Prepaid (insurance or subscriptions))/(Current liabilities)

Quick ratio = (Cash + Cash equivalents + Accounts receivable + Market securities)/(Current liabilities)

This quick ratio, at times referred to as the *acid-test ratio*, measures a company's ability to pay off short-term liabilities with assets on hand or current assets. A quick ratio of 0.95 means the quick assets are not enough to pay every dollar of current liabilities. A quick ratio of 1, however, indicates that a company's quick assets are equal to its current assets. This conveys that a company can pay off its current debts without selling its long-term assets. When a quick ratio is greater than 1, it means that the company owns more quick assets than current liabilities, and that is a good indicator. As the quick ratio gets larger, so does the liquidity of the company; for the most part, this is a positive sign indicating that more assets can quickly be converted to cash. This can demonstrate that the quick cash is reinvested into productive use. However, it should be determined if the reinvestment is going into nonproductive use instead. Quick assets are the sum of the cash, cash equivalent (like money markets accounts, certificates of deposits, saving accounts, treasury bills that mature within 90 days), and receivables of a company. It does not include other current assets such as inventory and prepaids (such as prepaid insurance), which can quickly turn into cash.

The current liabilities are obligations that must be paid within a year. They include interest on long-term debt that is paid within the next year. These liabilities may include taxes, wages, insurance, and utilities.

The quick ratio is a financial indicator of the ability to raise cash to pay bills due in the next 90 days. Quick ratios vary from industry to industry and may be seasonable. During difficult economic times a company may want to increase their quick ratio to deal with unforeseen shocks or turbulent times in the market. If a company has difficulty collecting accounts receivable, it may be good to increase cash so that you have money to cover the balance. If a company is growing, they may have a higher quick ratio in order to pay for investments and newly created investments.

If the quick ratio is 2.0, it means the company is able to quickly access its assets because they are two times the value of its short-term liabilities. Therefore, the company is liquid. However, if the quick ratio is 0.5, the company's quick ratio is half its short-term liabilities, and the company would not be able to cover its short-term liabilities. This ratio could be compared to other companies in the same sector to determine how it is doing.

The quick ratio is increasing, which also is a positive from 2xx0 (2.32) to 2xx1 (3.31). This quick ratio shows that the company is able to pay off its short-term liabilities quickly.

The cash ratio is a liquidity ratio that measures a company's ability to pay off short-term liabilities with highly liquid assets.

The cash ratio is a more conservative measure of the liquidity of a company position than the current ratio. A cash ratio of at least 0.5 to 1 is usually preferred for retail companies. However, cash ratios may not provide a good overall analysis of a company because it may be unrealistic for a company to hold large amounts of cash, particularly for startup, and small and medium-sized companies.

Cash ratio = (Operating cash flow)/(Current liabilities)

The cash ratio or cash asset ratio is a more conservative ratio where only cash and cash equivalents—the most liquid assets—are included in the calculation. Cash, for example, is cash, checking account, and bank drafts. Cash equivalents are assets that can be converted very quickly into cash, such as saving accounts, T-bills, and money market vehicles. Again, short-term liabilities include short-term debt, accrued liabilities, and accounts payable.

A ratio of 1 means that a company will be able to pay off the current liabilities with cash and cash equivalents with some funds left over. When the ratio is high, it indicates that the company can pay off its debt. However, a ratio too high may mean that the company is not using its assets right and investing in the company instead of letting the funds sit around. A ratio of 0.5 to 1 is preferred.

The cash ratio is increasing from 2xx0 (0.65) to 2xx1 (1.65). Since the cash ratio is even higher than the suggested rate of 1, the company is doing well even with the most conversative ratio for liquidity. However, the company needs to investigate if it should invest rather than just hold onto the money.

For the dress company, all of the KPIs at this point tend to suggest that the company is able to pay and repay its current loans and obligations.

Debt or Leverage Ratios

Leverage ratios are used to measure the debt level and obligations a company has. There are three leverage ratios discussed here: the debt ratio, interest ratio coverage, and debt service coverage. The debt ratio determines how much debt there is compared to the assets the company has. The interest ratio coverage shows if a company is able to pay off the interest expense given the EBITDA. The debt coverage indicates how much debt can be paid off with the company's operating income.

Debt ratio = (Total liabilities)/(Total assets)

This debt ratio is a company's amount of assets carried from their debt. And, if the company does hold debt, how does its credit financing compare to its assets? A higher ratio shows a higher rate of debt financing.

Lower ratios show better creditworthiness and are preferred, but too little debt has its own risks. A ratio of 0.4 or lower is considered better since the interest on a debt must be paid regardless of business profitability. A debt has interest risk and is interest risk sensitive. The higher the interest rate, the harder it is to pay the debt back. A company may need to declare bankruptcy if it cannot service its own debt. Therefore, a ratio higher than 0.6 may have trouble borrowing money.

Larger companies have more leverage with their negotiating lenders and are able to carry more debt than startup and small and midsize companies.

The debt ratio is decreasing from 2xx0 (0.38) to 2xx1 (0.27). Both these ratios are lower than 0.6, so if the company had to borrow money, it probably could at a low rate.

Interest coverage ratio = (Earnings before interest taxes, depreciation and amortization (EBITDA))/(Interest expenses) during a given period.

The interest coverage ratio measure shows how much a company can pay interest on outstanding debt. Many creditors use this measure to assess the risk of lending capital to the company. A higher ratio is better, and each industry may have an ideal ratio. The lower the ratio, the higher a company is burdened by debt. A ratio of less than 1 means that the company is not meeting its debt obligations. A ratio between 1 and 1.5 or lower may indicate that the company cannot meet its interest expenses.

Interest coverage is looked at over time to see if it shows a pattern of worsening, improving, or remaining stable.

The interest coverage increased from 2xx0 (28) to 2xx1 (36), which indicates that the company is able to cover its interest payments without an issue.

Debt service coverage ratio = (Net operating income)/(Total debt service)

This measure shows how fast the company's cash flow can pay off debt obligations. In other words, it indicates a company's ability to pay back loans. A ratio of 1 or above is considered a positive sign that a company is able to repay its debt and interest payments. A ratio of 2 or higher shows that a company is able to not only pay off their loans but also carry another loan if need be.

The debt service coverage ratio is from 2xx0 (3.5) to 2xx1 (4.8). These ratios are above 2 and show that the company is clearly able to cover their loan and interest payments.

From the leverage ratio, the dress company displays an ability to cover all its debt obligations.

Profitability Ratios

Profitability is important for all companies, but this book focuses on startups and small to medium-sized companies. Profit is incurred when expenses are subtracted from the revenue. Without profit, the company will not be able to secure any additional loans, attract investors, or grow. There are many different types of revenue KPIs that companies use to evaluate their businesses. There are six forms of revenue-generating KPIs the book discusses: profit margin, gross profit margin, net profit margin, EBITDA, operating profit, and return on assets. Profit margin is the percent profit made from the revenue generated. Gross profit margin is the revenue less the cost of goods (COGS) and a way to evaluate pricing accuracy. Net profit margin, or simply net margin, is the net sales over the revenue generated. Although gross profit margin, net profit margin, and operating are types of profits the company can produce, these metrics were explained in Chapter 4 and will not be reviewed here; they overlap both sections. This chapter discusses profit margin, EBITDA, and return on assets.

EBITDA is the overall ability of a company to generate profit from sales when fixed obligations such as taxes, depreciation, and amortization are considered. This is in the income statement. Return on assets shows how well a company turns its assets into money generated. All the profit metrics help determine how much money the company makes and, therefore, how well the company is managing its resources and how much profit the company could expect in the future.

Profit margin percent = (Revenue – Total expenses with the cost of goods)/ Revenue*100

The profit margin indicates how a company is handling its finances, and, more specifically, it compares the profit to sales. Profit margin tells the company how many cents per dollar the company generated for each dollar in sales. In general, a company with a profit margin percent over 20 percent is considered good and below 5 percent is not doing well, but this ratio can vary by industry and size of the company, as well as by other factors.

On the spreadsheet, the profit margin percent stayed stable from 2xx0 (70 percent) to 2xx1 (70 percent). The margin is remaining stable, and it shows that the company is making a marked profit. This profit margin percent is a bit high for a real company, but a more attainable profit margin percent is around 50 percent or more.

EBITDA = Net income – (Interest, depreciation, and amortization)

EBITDA means net income before interest, depreciation, or amortization.

EBITDA is a measure of a company's profitability or overall performance and may give a clearer view of a company's operations. EBITDA is used to measure a company's ability to generate a profit from sales. It is used sometimes in lieu of net income, also called *net profit*, and it does not include items of capital and financial expenditures, such as property, plants, and equipment. This metric adds back interest and tax expenses but excludes debt.

EBITDA is a measure that can be compared to other similar companies because it combines the core elements of a company's profit. A higher EBITDA is always better, and the number differs by sector, industry, and size of the company. Unfortunately, at times, companies use EBITDA to mask when the company has heavy debt and expensive assets.

The EBITDA increased from 2xx0 ($516,280) to 2xx1 ($625,036).

Return on assets = (Net income or profit)/(total assets)

Return on assets shows how well a company is relative to its assets. The net income comes from the income statement, and the total assets measure comes from the balance sheet. It gives the company owner a good idea of how well the assets generate earnings, in other words, how fast a product converts into cash and when the company can re-invest the cash. A return on asset over 5 percent is considered good.

The return on assets (ROA) increased from 2xx0 (47%) to 2xx1 (49%), and for both years these numbers are good.

Efficiency Ratios

Efficiency ratios are sometimes referred to as *activity ratios.* They give insight into how well the company manages its operations and sales activities. The goal of these activities is to produce income through the effective use of the company's resources.

Inventory turnover ratio = (Quantity sold)/Average inventory on hand

Inventory turnover reports how a company is able to sell its merchandise. Inventory turnover is important because one of the company's biggest expenses is inventory. Having too much inventory is costly because it takes up space and costs the company money to manufacture. The larger the inventory turnover ratio is, the faster a company needs to replenish its stock or else it will be at risk of running out of goods to sell.

Inventory turnover decreased from 2xx0 (2.63) to 2xx1 (2.3).

Inventory turnover demonstrates how many times a company has sold the merchandise and then replaced the inventory over a period of time. It is common to use sales or COGS, but sometimes it may be more useful to use units sold instead because it does not include markup costs or variable margins. Therefore, it may be a better predictor.

A high inventory turnover could indicate that the company has strong sales. However, it could also show that the company has insufficient inventory to cover all the sales. On the other hand, a low inventory turnover number could suggest that the company has excess inventory and weak sales. Therefore, this ratio will enable the company to decide on pricing, purchasing, and marketing of the product.

Average inventory = (Beginning inventory + Ending inventory)/2

The average inventory calculation for financial purposes is typically the inventory count at the beginning of the time frame and the end of the time frame, divided by 2. The average inventory for a time frame can also be the literal average of every inventory count available during the time frame, e.g., monthly, but sometimes these numbers can diverge.

The average inventory increases from 2xx0 (3,238 units) to 2xx1 (3,625 units), which implies a higher inventory liability for the time frame, and higher sales required to achieve strong turnover. Inventory liability is the risk of unsold merchandise via the cost of warehousing and marketing unsold merchandise until considered unsellable and disposed of, which also often involves a cost.

This fictitious company's financial metrics show that this company has enough liquidity to pay down its short-term debt and interest and loans. The company is stable and conservatively leveraged. Many new to midsize companies find themselves over-leveraged, and they need to watch this carefully so that they do not take on too much debt to sink the company altogether. Profit, which is an important component of a company, is stable and thereby allows the company to get competitive rates if it needs a loan. The company's inventory as well as the e-commerce department should be closely monitored to determine what the rate of each item turnover should be.

Summary

Analyzing financial ratios often is an essential part of any well-run and well-planned business practice. These metrics should be assessed frequently to ensure that the company can meet its short-term liquidity responsibilities to measure its accountability to pay off its debt and obligations owed. They are the most straightforward financial KPIs to demonstrate that a company is turning a profit and running an efficient business.

CHAPTER 6

Using Frequencies and Percentages to Create Stories from Charts

Charts should tell a visual story of the imperative aspects of a company's business questions including what the results and conclusions are and how the company's journey looks. Most managers and CEOs do not want to waste their time looking at endless data or complicated charts. Most senior executives want their employees to understand the business and what the relevant business decisions are enough to determine the pertinent aspects of the company and be able to display them in simple charts.

Charts can also be used to summarize data, see whether there are outliers, and determine if there are problems with the data and how to go about fixing them.

This chapter will review how to create simple pie charts and bar charts in Excel using frequencies and percentages. Although Excel is a powerful tool, its charts are basic; for more elaborate charts, use the open source software R or Python.

© Rhoda Okunev 2022
R. Okunev, *Analytics for Retail*, https://doi.org/10.1007/978-1-4842-7830-7_6

You saw real-life probability questions in Chapter 3, so now it is time to see how a small data set using two variables can be transformed into a visual representation. The first thing to do is to tally up and then summarize the data in a frequency table. We will use Excel, but this time we will let the data tell a simple story from each chart. However, before pictures are created, the formulas for frequencies and then percentages must be run and understood in order to know how to explain the business questions and the story's journey.

The book uses small databases to demonstrate the ideas; these techniques can calculate the formula and create charts but can easily be generalized and used with a larger database as well. In fact, small datasets should be used to test to make sure the data looks correct and you know how to use the techniques, but the real charts should utilize large datasets.

This chapter will first explain frequencies and percentages and then put them in charts.

Frequencies: How to Use Percentages

As shown in Figure 6-1, count each group's data for each month. Use frequencies for the histograms that have only a few data points in each group. It is not recommended to use percentages for small datasets or datasets that are very different in size because it may look like the business has more data than it really does.

	A	B	C	D
1				
2	Data			
3				
4		Subjects	Month	Shop
5			1= January	1= Store
6			2= February	2= Internet
7		1	1	1
8		2	1	1
9		3	1	1
10		4	1	1
11		5	1	1
12		6	1	2
13		7	1	2
14		8	1	2
15		9	1	2
16		10	1	2
17		11	1	2
18		12	1	2
19		13	1	2
20		14	1	2
21		15	1	2
22		16	2	1
23		17	2	1
24		18	2	1
25		19	2	1
26		20	2	2
27		21	2	2
28		22	2	2
29		23	2	2
30		24	2	2
31		25	2	2
32		26	2	2
33		27	2	2
34		28	2	2
35		29	2	2
36		30	2	2
37				
38				
39	Counting			
40				
41		Excel Code		
42	Subjects	Count	30	
43	1= January	CountIF	15	
44	2= February	CountIF	15	
45				
46				
47		Excel Code		
48	Shop	Count		30
49	1= Store	CountIF		9
50	2= Online	CountIF		21
51				
52				

Figure 6-1. *Data and count of the month and shop mode*

	A	B	C	D
38				
39	*Counting*			
40				
41		Excel Code		
42	Subjects	Count	=COUNT(C7:C36)	
43	1= January	CountIF	=COUNTIF(C7:C36,1)	
44	2= February	CountIF	=COUNTIF(C7:C36,2)	
45				
46				
47		Excel Code		
48	Shop	Count		=COUNT(D7:D36)
49	1= Store	CountIF		=COUNTIF(D7:D36,1)
50	2= Online	CountIF		=COUNTIF(D7:D36,2)
51				

Figure 6-1. (*continued*)

The example scenario is that a small business is having an email campaign for loyal customers and wants to know, based on subjects for two months (January and February), which method of purchase is used more often. The method of purchase is referred to as the *store* (in-store or Internet), which is the type of buyers' shopping patterns. Both January and February have 15 buyers, so there are 30 shoppers in total. Nine shopped in-store, and 21 on the Internet. Half of the shoppers shopped in January, and the other half shopped in February. Shoppers were equally likely to shop in January as in February. The difference in size of these groups and the fact that these are small groups may make it hard to compare shoppers who shop in the store compared to those online. Figure 6-1 displays the data and Excel code to tally and calculate the frequency for each group.

Total number of subjects = 30

Total number of 1 (Month = January) = 15 subjects

Total number of 2 (Month = February) = 15 subjects

Total Shop 1 (Shop = Store) = 9 subjects

Total Shop 2 (Shop = Online) = 21 subjects

The next step is to calculate the percentage of shoppers in January. Simply take the count of shoppers in January and divide it by the count in the entire sample and multiply that number by 100. Again, this is the number of successes divided by the entire group in the sample. Each month has the same number of shoppers, and 50 percent of the shoppers are in January and 50 percent are in February. As percentages, they always need to add up to 100 percent. The cumulative percent also adds up to 100 percent when all numbers are added together, as shown in Figure 6-2.

	Interval	Frequency	%	cum %
January	1	15	(15/30) X100=50%	50%
February	2	15	(15/30) X100=50%	50%+50%=100%
Total		30	100%	

	Interval	Frequency	%	cum %
Store	1	9	(9/30)*100=30%	30%
Internet	2	21	(21/30)*100=70%	30%+70%=100%
Total		30	100%	

Figure 6-2. *Calculating the number of shoppers in January and February*

Once the frequency for store and Internet are tallied, take the total number in the store, which is 9, and divide it by the total number in the sample, which is 30; then multiply that number by 100. Here, the percentages of 30 percent and 70 percent are calculated, respectively. Again, the cumulative percentage is initially 30 percent for stores but then add 30 percent and 70 percent for the store and Internet. Always remember that a percent is between 0 percent and 100 percent, but always sum up to 100 percent. Remember, round shoppers to an integer because there is either a person or not.

The percent column has to add up to 100 percent, while in the cumulative percent column the individual percentages add up one at a time until all the percentages are included, which will include 100 percent of the data. If this does not happen, then there is a calculation error. If the calculation is very close to 100 percent, then there may be a rounding error.

Figures 6-3 and 6-4 illustrate the page percentages and frequencies in Excel.

	A	B	C	D	E	F
1						
2						
3	Percents and Cumulative Percents					
4						
5						
6		Month	Interval	Frequency	%	cum %
7		January	1	15	0.50	0.50
8		February	2	15	0.50	1.00
9						
10			Excel code			
11		Total	Sum	30	1.00	
12						
13						
14						
15		Shop	Interval	Frequency	%	cum %
16		Store	1	9	0.30	0.30
17		Internet	2	21	0.70	1.00
18						
19			Excel code			
20		Total	Sum	30	1.00	
21						

Figure 6-3. Frequencies, percent, and cumulative percent in Excel

	A	B	C	D	E	F
1						
2						
3	Percents and Cumulat					
4						
5						
6		Month	Interval	Frequency	%	cum %
7		January	1	15	=D7/$D$11	=E7
8		February	2	15	=D8/$D$11	=F7+E8
9						
10			Excel code			
11		Total	Sum	=SUM(D7:D8)	=SUM(E7:E8)	
12						
13						
14						
15		Shop	Interval	Frequency	%	cum %
16		Store	1	9	=D16/$D$11	=E16
17		Internet	2	21	=D17/$D$11	=F16+E17
18						
19			Excel code			
20		Total	Sum	=SUM(D16:D17)	=SUM(E16:E17)	
21						

Figure 6-4. *Excel code of frequencies, percent, and cumulative percent*

Although frequencies can explain data, they are not always so clear to the viewer. Horizontal, vertical, and pie charts explain data as well and can tell a story in just a glance. Let's review them now.

Simple Charts: Horizontal, Vertical, and Pie

To create horizontal, vertical, or pie charts, go to the Insert tab in Excel and review the percentages you created from the raw data. There you will find a variety of bar charts and pie charts. Excel also has a recommendation icon that can help you to figure out which type of chart to use. Although there are no absolute rules for which chart to use, a key to remember is that each chart should be as simple as possible and present a clear and precise statement about your findings; it should tell the story that is important for the business to understand.

In the following case, the horizontal and vertical charts and pie chart are used to tell an understandable and uncomplicated story. The pictures of the pie chart may display the most poignant and effective chart. Let's see why, but first let's go over the data and the storyline questions.

For the most part, most of the charts that will be generated to tell the company story will be from the statistically significant results.

As can be seen from the data, the number of shoppers for the variables Month and Store are very different. For Store, the count of data points for in-store is 9, and for Internet the count of data points is 21. Again, the book is demonstrating these concepts using small groups, but samples 30 or more should be used, or a more sophisticated statistical analysis determining the sample size should be used. Also, the sizes of these two responses are disparate, and both samples are smaller than 30. Caution should be used when comparing these groups because having very different size datasets can skew the results and lead to spurious conclusions. In such cases, it may be better to just state the probabilities with their associated numbers without claiming statistical significance.

For the shoppers for the Month variable, both January (15) and February (15) have the same small count or number of subjects. In this case, an independent t-test can be used to compare the data, and this statistical test will be explained in a later chapter.

Here are some ideas that may be derived from the data to create a story. There is a larger percentage of the shoppers from a store shop on the Internet (70 percent) than in the store (30 percent). Loyal customers buy about the same whether it is January or February (50 percent). Are these results significant? We could have added to the story by including what it would say about revenue generated.

The next part is to create the type of chart that displays the story you are trying to tell. It is important to look at each and see which type of chart would accurately and simply show the story you are trying to tell at a glance.

Horizontal and Vertical Bar Charts

Figures 6-5 and 6-6 are examples of horizontal bar charts.

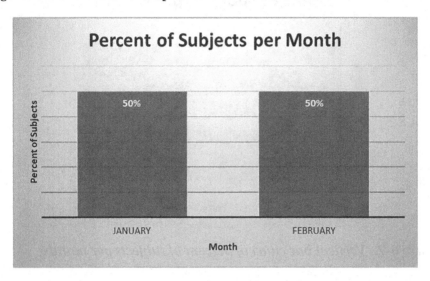

Figure 6-5. *Horizontal bar chart of percent of subjects per month*

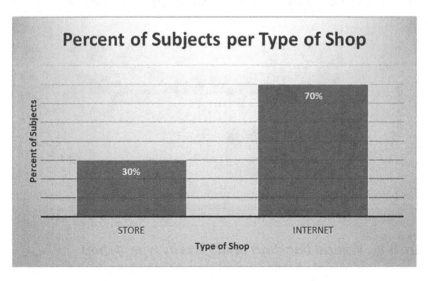

Figure 6-6. *Horizontal bar chart of percent of subjects by type of shop*

Figures 6-7 and 6-8 are examples of vertical charts.

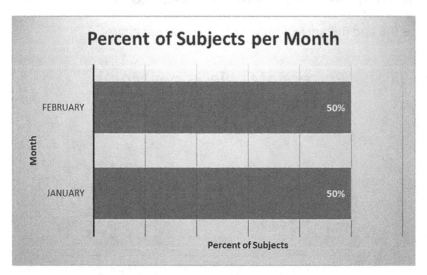

Figure 6-7. *Vertical bar chart of percent of subjects per month*

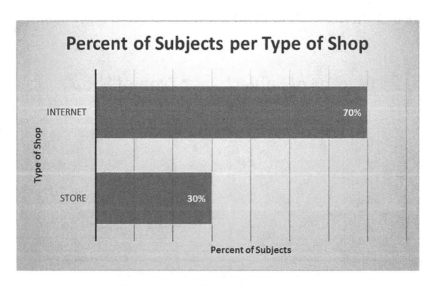

Figure 6-8. *Vertical bar chart of subjects by type of shop*

Pie Charts

A pie chart is another type of chart used to show numbers and percentages. For instance, with shops, the pie chart in Excel shows immediately that 50 percent of the data is in January and 50 percent is in February and that each group has 15 in each (Figure 6-9). Here, the researcher can identify if there are any numbers that do not belong in the database.

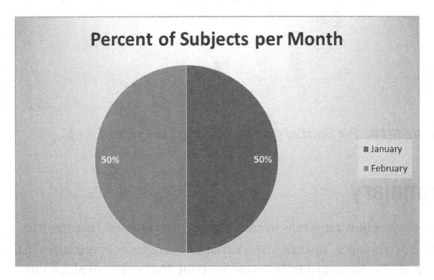

Figure 6-9. *Pie chart of percent of subject per month*

Figure 6-10 clearly implies that it may be difficult to compare Internet shoppers to store shoppers because of the difference in the size of the groups. And these charts could be misleading and should be used only with caution. Most stores have large enough datasets to make this type of chart usable, so I display it here to show how to use it.

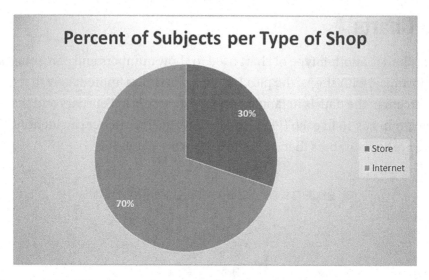

Figure 6-10. *Pie chart of percent of subject per type of shop*

Summary

This chapter illustrates how to create frequency tables and use them to create percentages. You can insert either frequencies or percentages in the charts depending on which is more powerful. If there is a small sample, then numbers are recommended. At that point, you can decide which types of chart should be used and if the data warrants recognition enough for the employee to create charts from it. A simple and clear chart can tell a story at a glance in a way that a frequency table or data never could.

To elaborate and tell a more elaborate story about a business problem, the reader may want to determine if these results are significant. In this case, what would the chart say about how much revenue is generated in a store or online senario?

CHAPTER 7

Hypothesis Testing and Interpretation of Results

Hypothesis tests define why you are conducting statistical analysis and what particular meaning the results have for the study being conducted for your company. The findings of the statistical tests answer the business questions that are being researched. However, for this to be accomplished, the objectives of the study need to be clearly defined.

This cannot be emphasized enough. The researcher needs to understand what the storyline is for their research at hand. In other words, what are the business questions, and why are the researchers trying to find answers to those questions in the order and manner that they are doing it? What is the storyline of the journey the business is on, and why is the business trying to find those answers? That is what needs to be defined and delineated. Clear objectives about the motivation of the storyline need to be defined prior to the analysis so that the appropriate data can be collected and the data can be analyzed in the right way with the relevant statistical methods. The work that is done prior to collecting the data and conducting the statistical analysis is the crux of whether the results make any sense at all.

I have seen cases where variables were randomly chosen, the data was collected, and the data was cleaned and scrubbed carefully and then brought to a statistician to be analyzed; however, the business questions

© Rhoda Okunev 2022
R. Okunev, *Analytics for Retail*, https://doi.org/10.1007/978-1-4842-7830-7_7

were not asked prior to collecting the data. The statistician could not do anything with the data, and the data was deemed worthless, because very few of the business objectives were able to be defined and explained. All that work of collecting the data and no clear and coherent story could be told from those results. How disappointing is that?

Once the business decisions are made, the data variables are chosen, the data is collected and cleaned, and all outliers and missing data is dealt with in a suitable and consistent manner, then it is time to conduct the statistical analysis with the clearly outlined hypotheses.

When setting up a hypothesis, keep in mind that as with a court of law, a hypothesis that is not significant does not say that a person is innocent or not guilty of committing a consequential action; it says that there is not enough evidence to prove that the person did not brake the law. It does define whether a person is found guilty of a crime.

Scientifically, a null hypothesis states that there is not enough evidence to prove that there is a difference or an effect on the outcome or that there are statistically significant results that do have a noteworthy effect or change on the outcome. An alternative hypothesis states that there is a difference or an effect on an outcome or there is a statistically significant result that do have noteworthy effect or change on the outcome.

There are five steps to follow to determine whether a statistical test is significant, determine whether the results are meaningful, state why the results have a meaning, and determine what that meaning is.

Step 1: The Hypothesis, or Reason for the Business Question

The first step needed with a statistical procedure is to understand what the study's hypothesis is investigating. You need to have a clear idea of what the project is about and why you are conducting it; clearly know your research questions. The null hypothesis (H_0) states that there is no change in the variables or that there is not enough evidence to reject H_0. On the other hand, the alternative hypothesis (H_1) states that a change or effect in variables occurred and the results are significant. The alternative hypothesis says the change in conditions does affect the results.

A two-tailed test is used here and is the most common statistical method used. For instance, in an Independent t-test using a two-tailed test the null hypothesis uses the two population parameters mu one and mu two, the population means for the two groups, and they are set equal to each other. For the alternative hypothesis for a two-tailed test the two population parameters for mu, of the population means, are set not equal to each other. (For the alternative hypothesis of a one-tailed test one mean mu would be greater than or less than the other one). Although there is some debate in the literature about which hypothesis test to use—a one-tailed test or a two-tailed test—the two-tailed test is the standard. This book will discuss only the two-tailed test because it is the most widely used. A two-tailed test is usually used unless the researcher clearly understands the direction of the hypothesis and knows how to design the research as such. If a two-tailed test is not used and the question is asked in the wrong direction, the results will be totally wrong, and the investigator will miss evaluating the hypothesis being posed.

Therefore, a good practice is to use a two-tailed test unless the researcher clearly knows the direction, in which case a one-tailed test can be used. Also, one-tailed tests are significant more often than two-tailed tests.

Step 2: Confidence Level

The second step is determining the level of confidence used. The confidence coefficient (1 - alpha) is the probability that the null hypothesis is true, or that there is not enough evidence to reject the null hypothesis.

The confidence level is (1 - alpha) * 100%. Alpha is the probability that the null hypothesis is rejected given that the null hypothesis is true or a type one error. Since alpha is selected before a statistics test is conducted the risk of performing a type one error is controlled. And, alpha is referred to as the level of significance. The rejection region, area or critical region is determined by the alpha level. Often the alpha is 0.05. Although other alpha can be used, such as 0.01 or 0.10, the default alpha in Excel and other statistical software is 0.05. The confidence level is 95%, 99% and 90%. The p-value is compared to the alpha level in statistical tests. The p-value is the probability of a probability distribution, such as the Normal Curve, determined by a statistical test results and is shown and will be reviewed on all the excel outputs. If the p-value is greater than or equal to the alpha, there is not enough evidence to reject the null hypothesis. If the p-value is less than alpha, then reject the null hypothesis.

Since a small sample is used, knowing and calculating the degrees of freedom for each statistic where it is necessary is a must.

Step 3: Mathematical Operations and Statistical Formulas

The third step is using the statistical test that was decided on when organizing the data with the research questions in mind and then performing the mathematical operations for those statistical tests. In other words, the data is run through the formula, and all the results are calculated. The statistical tests are essential to know ahead of time in order to have the data configured so that those procedures can be used.

Step 4: Results

The fourth step is stating the results in the order that the results will have on your business. The results need to be laid out so that the story can be told. The next two chapters cover two different types of statistical procedures:, the Pearson correlation and the independent t-test, which will further expand upon this section. Appendix A reviews the different data types that are used for t-tests and Pearson correlations.

Step 5: Descriptive Analysis

The fifth step is elaborating on the results and explaining the conclusions for future studies. This is where graphs are important to use to demonstrate the points of the results and to start putting together the story of the statistics.

Summary

All five steps are needed to perform a hypothesis test for different statistics. It cannot be stressed enough how important it is to understand the business question at the beginning of the investigation and to determine the variables that will help answer the research questions before the investigation has begun because this will help with knowing the statistical methods that will be involved. Once the analysis has been conducted, the researcher will notice other areas or questions that they may have for further investigation. In the next chapter, these five steps are used for both a Pearson correlation and independent t-tests.

Pearson Correlation and Using the Excel Linear Trend Equation and Excel Regression Output

Chapter 7 introduced the idea of thinking through a business question and building a hypothesis test around that query, and it elucidated on the five steps to run a statistical inquiry. This chapter shows how to implement those five steps using the statistical test called the *Pearson correlation*. A small dataset is used for demonstration purposes to show how to use the Pearson correlation, also referred to as *Pearson's r correlation*.

Pearson Correlation Defined

A Pearson or simple correlation is a parametric statistical test that shows a linear relationship between two continuous random variables. The data needs to be randomly distributed from a large enough sample or taken from a normal population. For example, the analyst may suspect that there

© Rhoda Okunev 2022
R. Okunev, *Analytics for Retail*, https://doi.org/10.1007/978-1-4842-7830-7_8

is a positive linear relationship between the number of items sold and the total revenue generated from those purchases but wants to verify whether this trend is significant. Just because there is a relationship does not mean the variables are causally related. The investigator first needs to make sure that the business questions that are being researched make sense and are meaningful and substantial. For instance, an experiment may find a positive relationship between subjects who eat more chocolate and are taller. This, of course, does not mean very much and does not make sense and should not be used. Therefore, correlations do not necessarily mean that one causes the other. When devising this analysis, remember to focus on variables that are significant and important to your investigation. For instance, are the number of items sold and total revenue positively correlated? These two variables are related in a retail setting, and a store owner wants to find a positive relationship between these two variables. That is, as the number of items sold increases, so does the revenue. This finding would be meaningful and valuable to your company. If the researcher does not find a positive relationship between these variables, it is important to try to think why this occurrence happened; it may be because there were a lot of returns, for example. Do not necessarily blame it on bad data or a computer glitch.

A Pearson correlation has three basic characteristics. The first is the direction, which could be positive or negative. A positive correlation could go to +1, which is a perfect positive correlation. Thus, as the X-variable increases, the Y-variable increases. A negative correlation could go to -1 and means that as the X-variable increases, the Y-variable decreases. A zero correlation means no correlation or relationship exists between the variables. This result is a straight horizontal line across. The second characteristic is that the strength extends from -1 through 0 to +1. Third, the form of a Pearson correlation is linear.

In sum, a Pearson correlation measures the direction, linearity, and strength of an association between two continuous variables to determine if the relationship is statistically significant. The letter r represents a Pearson correlation. The formula used in this book for a Pearson correlation shows X and Y vary together in the numerator of the equation and vary separately in the denominator of the equation. This makes it easy to understand how Pearson correlations work.

Hypothesis Testing and Descriptive Steps for a Pearson Correlation

Now, let's apply the five steps of hypothesis testing to a retail example. An independent variable X and an outcome or dependent variable Y are defined. In this case, the variables related are (X) number of items bought and (Y) total revenue received from the sales.

Hypothesis:

As explained previously, a two-tailed test is used. H_0 explains that there is not enough evidence to suggest there is a linear relationship, while H_1 says there is a significant relationship between the two variables. Again, the hypothesis direction of the significance is not being tested, and it is a two-tailed test.

Step 1: The Hypothesis, or the Reason for the Business Question

p is pronounced "rho" and is a Greek letter used in the hypothesis for the Pearson correlation.

$$H_0: p = 0$$
$$H_1: p \neq 0$$

The null hypothesis means that there is not enough evidence to assume that there is a relationship between the two variables. In this case, it would mean that there is not enough evidence to assume that the items bought are correlated with the total revenue. In other words, the correlation is zero or close to it.

The alternative hypothesis means that there is a significant relationship between the two variables—items bought and total revenue—but it does not tell you the direction or the relationship. The two-tailed test is used when the null hypothesis uses the notation of "equal to" and the alternative uses the notation of "not equal to." The two-tailed test is the standard or the default analysis on most statistical packages including Excel. The direction can be seen in the graph trend as to whether it is in the positive direction, negative direction, or zero, which is a straight line across.

Step 2: Confidence Level

The Pearson correlation uses degrees of freedom equal to df = n-2. This number 2 is used because two points make up a line. Degrees of freedom (df) is the number of values in a calculation that have the freedom to vary.

Therefore, df = n-2, and in this case, n is the number of total observations, which is 9 so that 9–2 = 7. Seven corresponds with the df on the output sheet in the result section.

The confidence level is 95%. Alpha or the significance level = 0.05 is used for the rejection or critical region, which is the default in statistical packages including Excel.

The data used in the Pearson correlation data for the two variables should be normally distributed, and each group should be 30 or more. Again, a statistician can help with determining a better size sample to use. The data needs to be from a normally distributed population. Linearity assumes that the data will form a linear trend.

Step 3: Mathematical Operations and Statistical Formula

The researcher is determining whether there is a relationship between the number of items bought (X = independent variable) and revenue generated (Y = dependent variable). X is the independent variable that is the variable that is the cause of the study and can be varied, and Y is the dependent variable that is the variable tested and measured. In this case, "items bought" is the independent or input variable because the customer is deciding how many items to buy and the number of items bought varies per customer. On the other hand, the dependent variable is the revenue generated because the amount spent depends on the items bought.

To calculate the Pearson correlation formula (r = SP / sqrt (SSX times SSY)) and the simple linear regression equation (Y-hat = mX +b), the following steps must be performed:

1. Multiply X against itself, do the same for the Y, and multiply X and Y together, as shown in Table 8-1. Then sum the columns for X, Y, XX, YY, XY, and Y-hat (a *hat* is an estimate) and calculate the mean of X or X-bar and Y or Y-bar. The Y-hat is an estimate because it is the best-fit regression line that uses the least square technique. The Y-hat is the predicted value in the regression and will be used in Chapter 11, the Forcasting Chapter.

Table 8-1. *Summary Output for the Pearson Correlation*

X Items Bought	Y Total Revenue	XX	YY	XY	Y-hat= mX + b
5	25	25	625	125	77
10	250	100	62,500	2,500	319
2	10	4	100	20	(67)
15	900	225	810,000	13,500	560
20	1,000	400	1,000,000	20,000	801
17	300	289	90,000	5,100	657
5	68	25	4,624	340	77
20	700	400	490,000	14,000	801
5	50	25	2,500	250	77
99	3,303	1,493	2,460,349	55,835	Total
11	367				Mean

2. Calculate the totals of all the columns needed.
   Use the following formulas and put in the totals
   calculated earlier to derive the results:

   $$SP = \Sigma\,(X - X\text{-bar})\,(Y - Y\text{-bar}) = \Sigma XY - (\Sigma X \Sigma Y)/n = 19{,}502$$
   $$SSX = \Sigma\,(X - X\text{-bar})\,(X - X\text{-bar}) = \Sigma XX - (\Sigma X \Sigma X)/n = 404$$
   $$SSY = \Sigma\,(Y - Y\text{-bar})\,(Y - Y\text{-bar}) = \Sigma YY - (\Sigma Y \Sigma Y)/n = 1{,}248{,}148$$

3.   Take the results and put them into the formula for r, as
shown next. This is the Pearson correlation formula.

$$r = SP / sqrt (SSX \times SSY) = 0.8684$$

To calculate the simple regression equation, both m and b need to be
calculated as follows:

$$m = SP/SSX = 48.272$$
$$b = mean(y) - mean(x) \times b = -163.995$$
Y-hat = b + mX Regression equation Y-hat = intercept + slope * X-variable
$$Y\text{-hat} = -163.995 + 48.272 \times X$$

The regression equation identifies the likelihood of predicting future
outcomes based on the historical sales data accumulated. The purpose is to
optimize a prediction or scientifically create a best assessment of what will
happen in the future. The direction of the slope is always the direction of the
linear trend. If the slope is positive there is a positive trend and if the slope is
negative there is a negative trend. Therefore, what does the predictive Y-hat
equation mean, and how should the company plan in the future?

The X is the independent or explanatory variable. The Y is the
dependent or response variable. The Y-hat is the predicted value in the
regression equation, and it is the linear regression equation that best fits
for the data. The m is the slope of the equation, and the b is the intercept
for the line. The m is often referred to as b_1 and the b as b_0 in regression
equation books. To make it more readable and understandable that it is
a linear line m and b are used in this book. The last part is to calculate
r-squared, which is just r*r. Since you multiply it by itself, the number will
always be positive. Excel refers to the Pearson correlation as multiple r.

Total Variation = Explained variation + Unexplained variation

$$\Sigma(Y_i - Y\text{-hat})^2 = \Sigma(Y\text{-hat}_i - Y\text{-bar})^2 + \Sigma(Y_i - Y\text{-hat}_i)^2$$

In other words, r-squared is the percentage of the explained variation over the total variation.

Using what was just learned, now it is time to put it to use in order to explain what has been done.

Step 4: Results

Since the results from the Pearson statistics are 0.8685, which has a p-value of 0.002 that is less than an alpha value of 0.05, the result is significant, and there is enough evidence at the 0.05 alpha level to show that there is a significant correlation or relationship between the two variables. Also, 0.8685 is close to 1.

When the Data Analysis ToolPak in Excel is used for the regression, it shows the results in Figure 8-1.

SUMMAY
OUTPUT

Regression Statistics	
Multiple R	0.868471212
R Square	0.754242246
Adjusted R Square	0.719133995
Standard Error	209.3329711
Observations	9

ANOVA

	df	SS	MS	F	Significance F
Regression	1	941405.9505	941405.9505	21.48333319	0.002383564
Residual	7	306742.0495	43820.29279		
Total	8	1248148			

	Coefficients	Standard Error	t Stat	P-value	Lower 95%	Upper 95%	Lower 95.0%	Upper 95.0%
Intercept	-163.9950495	134.1391667	-1.22257394	0.261052528	-481.1837762	153.1936772	-481.1837762	153.1936772
X Items Bought	48.27227723	10.41470457	4.635011671	0.002383564	23.64541423	72.89914022	23.64541423	72.89914022

Figure 8-1. *Output of the regression analysis*

The multiple r is the Pearson correlation, which is 0.8685 in Excel. The Pearson correlation is significant because the variable X, which represents the items bought, has a P-value of 0.00238 and because it is smaller than the alpha level of 0.05. Therefore, there is a significant relationship between items bought and total revenue.

Step 5: Descriptive Analysis Interpretation of Results

The graph in Figure 8-2 can be created when using the scatter plot graph on the Insert tab with the default linear trendline, which is the linear regression line. The scatter plot demonstrates pictorially the direction of the data, which is positive.

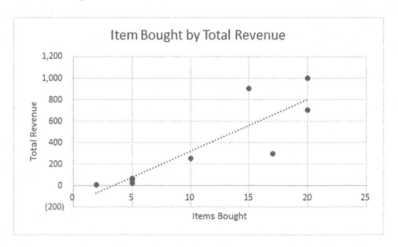

Figure 8-2. *Correlation between items bought and total revenue*

Since the results are significant, this means there is statistically significant evidence to say that the more items you buy, the higher the bill will be, or there is a positive correlation between the items bought and total sales.

This line is called the regression line or *least square line*; it is the best-fit line, and it minimizes the variance. It also predicts what a Y-hat value will be given an X value. The least square line is the trendline in Excel, and it is the best-fit line for the data that is used.

Three Examples Using Small Datasets

To make sure you clearly understand Pearson correlations, this section will give a negative, a positive, and no correlation example with small datasets. As mentioned previously in the book, small datasets are for demonstration purposes only to give you an understanding of the results of the findings.

These examples have four subjects, and only the Y or dependent variable changes to clearly show how correlations work for the negative, positive, and no correlation results.

Step 1: Hypotheses Are All the Same

The following hypotheses will be the same for all the examples:

$$H_0: p = 0$$
$$H_1: p \neq 0$$

All the examples are two-tailed.

Step 2: Level of Confidence

Since df = 2, the alpha level will be 0.05.

$$df = n-2 = 4-2 = 2$$

Step 3: Mathematical Operations and Statistical Formula

At first we will calculate the equations by hand and then run the procedure. The diagrams are shown in this section to illustrate the direction of the predicted line. We could have presented this material in step 5, but it will be easier to understand here.

Negative Correlation

There is a negative correlation when the X (independent variable) and Y (dependent variable) go in opposite directions (Figures 8-3 and 8-4). This is shown in the graph in Figure 8-5 after the formula calculations, so you can clearly see that there is a negative correlation and the slope is negative.

	X	Y	XX	YY	XY	Y-hat
	70	4	4900	16	280	4.6
	60	8	3600	64	480	7.2
	50	10	2500	100	500	9.8
	40	12	1600	144	480	12.4
Totals	220	34	12600	324	1740	
Mean	55	8.5				
n	4					

Figure 8-3. *Calculation conducted for the Pearson correlation*

94

X - Mean(X)	(X-mean(X))^2	Y - Mean(Y)	(Y- Mean(Y))^2	(X-M(X))*(Y- M(Y))
15	225	-4.5	20.25	-67.5
5	25	-0.5	0.25	-2.5
-5	25	1.5	2.25	-7.5
-15	225	3.5	12.2	-52.5
Totals 0	500	0	35	-130

Figure 8-4. *Summary of the calculations of a negative pearson correlation*

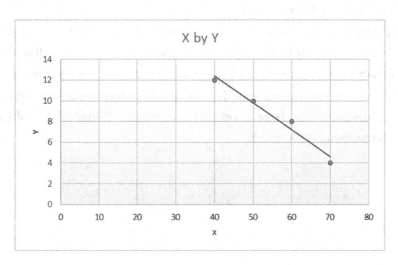

Figure 8-5. *Linear trend of a negative Pearson correlation*

Here are the steps to calculate the r correlation and regression line:

$$SP = (X-M(X))*(Y-M(Y)) = \Sigma XY - (\Sigma X \Sigma Y)/n = -130$$
$$SSX = (X-mean(X))^2 = \Sigma XX - (\Sigma X \Sigma X)/n = 500$$
$$SSY = (Y-mean(Y))^2 = \Sigma YY - (\Sigma Y \Sigma Y)/n = 35$$
$$r = SP/(sqrt(SSX*SSY) = -0.9827$$
$$b = mean(Y)-mean(X) = 22.8$$
$$m = SP/SSX = -0.26$$
$$Y = b + mX = 22.8 - 0.26X$$

In this case, the m, which is the slope, is negative, and therefore the linear trend is negative. This shows that as the X variable increases, the associated Y variable decreases.

In Figure 8-6, the multiple R is 0.982 out of 1 in the negative direction, so it is -0.982. The graph also shows the negative direction, as well as that the X variable goes up while the Y variable goes down. The P-value for the Pearson Correlation test is 0.017, which is less than the alpha of 0.05. This demonstrates that there is a significant negative linear relationship between the variables X and Y.

SUMMARY
OUTPUT

Regression Statistics

Multiple R	0.98270763
R Square	0.965714286
Adjusted R Square	0.948571429
Standard Error	0.774596669
Observations	4

ANOVA

	df	SS	MS	F	Significance F
Regression	1	33.8	33.8	56.33333333	0.0172924
Residual	2	1.2	0.6		
Total	3	35			

	Coefficients	Standard Error	t Stat	P-value	Lower 95%	Upper 95%	Lower 95.0%	Upper 95.0%
Intercept	22.8	1.94422221	11.7270546	0.007193106	14.434687	31.165313	14.434687	31.165313
X	-0.26	0.034641016	-7.505553499	0.01729237	-0.4090483	-0.110951737	-0.4090483	-0.110951737

Figure 8-6. *Output of an example of a negative Pearson correlation*

97

Positive Correlation

Notice in Figure 8-7 that the Xs (independent variables) and the Ys
(dependent variable) move in the same direction. For both the Xs and Ys,
the numbers are increasing.

	X	Y	XX	YY	XY	Y-hat
	70	12	4900	144	840	12.4
	60	10	3600	100	600	9.8
	50	8	2500	64	400	7.2
	40	4	1600	16	160	4.6
Totals	220	34	12600	324	2000	
Mean	55	8.5				
n	4					

Figure 8-7. *Calculations conducted for the Pearson correlation*

Notice in Figure 8-8 that (x − mean(x)) and (y − mean(y)) both equal 0.
That is a demonstration of reversion to the mean.

X - Mean(X)	(X-mean(X))²	Y - Mean(Y)	(Y-Mean(Y))²	(X-M(X))*(Y-M(Y))
15	225	3.5	12.25	52.5
5	25	1.5	2.25	7.5
-5	25	-0.5	0.25	2.5
-15	225	-4.5	20.25	67.5
Totals 0	500	0	35	130

Figure 8-8. *Summary of the calculations of a positive Pearson correlation*

Here are the steps to calculate the r correlation and regression line:

$$Mean = M$$
$$SP = X-M(X)*Y-M(Y) = \Sigma XY -(\Sigma X \Sigma Y)/n = 130$$
$$SSX = (X-mean(X))^2 = \Sigma XX - (\Sigma X \Sigma X)/n = 500$$
$$SSY = (Y-mean(Y))^2 = \Sigma YY - (\Sigma Y \Sigma Y)/n = 35$$
$$r = SP/(sqrt(SSX*SSY) = 0.9827$$
$$b = mean(Y)-mean(X) = -5.8$$
$$m = SP/SSX = 0.26$$

In this case, the m, which is the slope, is positive, and therefore the linear trend is positive. This shows that as the X variable increases, the associated Y variables increases.

In Figure 8-9, the multiple R is 0.982 out of 1 in the positive direction, so it is 0.982. The positive direction is also shown in the graph, as well as that the X variable goes up while the Y variable goes up. The P-value for the Pearson Statistical Test result is 0.017, which is less than the alpha of 0.05. This demonstrates that there is a significant positive linear relationship between the variables X and Y.

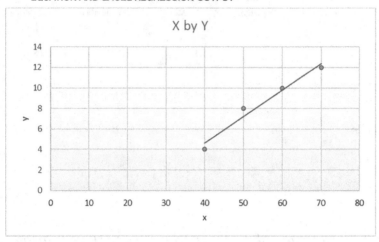

Figure 8-9. *Linear trend of a positive Pearson correlation*

In Figure 8-10, notice that this time there is a positive correlation; when the data for the X variable gets larger, so does the Y variable. This time the correlation is positive 0.982 out of 1 because it goes in the positive direction.

$$Y = b + mX = -5.8 + 0.26X$$

SUMMARY OUTPUT

Regression Statistics	
Multiple R	0.98270763
R Square	0.965714286
Adjusted R Square	0.948571429
Standard Error	0.774596669
Observations	4

ANOVA

	df	SS	MS	F	Significance F
Regression	1	33.8	33.8	56.33333333	0.01729237
Residual	2	1.2	0.6		
Total	3	35			

	Coefficients	Standard Error	t Stat	P-value	Lower 95%	Upper 95%	Lower 95.0%	Upper 95.0%
Intercept	-5.8	1.94422221	-2.983198099	0.096393418	-14.165313	2.565312997	-14.165313	2.565312997
X	0.26	0.034641016	7.505553499	0.01729237	0.110951737	0.409048263	0.110951737	0.409048263

Figure 8-10. *Output of an example of a positive Pearson correlation*

In this case, the m, which is the slope, is positive, and therefore the
linear trend is positive. This shows that as the X variable increases so does
the associated Y variable.

No Correlation

Notice in Figure 8-11 that the Xs and the Ys do not move in any meaningful
pattern. Therefore, there is no correlation or relation. A straight line across
or close to a straight line across the graph will appear.

X	Y	XX	XX	YY	XY	Y-hat
70	8	4900	4900	64	560	8.8
60	12	3600	3600	144	720	8.6
50	4	2500	2500	16	200	8.4
40	10	1600	1600	100	400	8.2
Totals	220	34	12600	324	1880	
Mean	55	8.5				
n	4					

Figure 8-11. *Calculations conducted for the Pearson correlation*

Figure 8-12 summarizes the calculations for no correlation using the
Pearson correlation.

X - Mean(X)	(X- mean(X))²	Y - Mean(Y)	(Y- Mean(Y))²	(X-M(X))*(Y- M(Y))
15	225	3.5	12.25	-7.5
5	25	1.5	2.25	17.5
-5	25	- 0.5	0.25	22.5
-15	225	- 4.5	20.25	-22.5
Totals 0	500	0	35	10

Figure 8-12. *Calculations for no correlation using the Pearson correlation*

Here are the steps to calculate the r correlation and regression line:

$$SP = X-M(X)*Y-M(Y) = \Sigma XY -(\Sigma X \Sigma Y)/n = 10$$
$$SSX = (X-mean(X))^2 = \Sigma XX - (\Sigma X \Sigma X)/n = 500$$
$$SSY = (Y-mean(Y))^2 = \Sigma YY - (\Sigma Y \Sigma Y)/n = 35$$
$$r = SP/(sqrt(SSX*SSY) = 0.0756$$
$$b = mean(Y)-mean(X) = 7.4$$
$$m = SP/SSX = 0.02$$

In Figure 8-13, the m, which is the slope, is close to zero, and therefore there is not a significant linear trend. This shows a straight line across the x-axis and no significant linear trend.

SUMMARY OUTPUT

Regression Statistics	
Multiple R	0.075592895
R Square	0.005714286
Adjusted R Square	-0.491428571
Standard Error	4.171330723
Observations	4

ANOVA

	df	SS	MS	F	Significance F			
Regression	1	0.2	0.2	0.011494253	0.924407105			
Residual	2	34.8	17.4					
Total	3	35						

	Coefficients	Standard Error	t Stat	P-value	Lower 95%	Upper 95%	Lower 95.0%	Upper 95.0%
Intercept	7.4	10.46995702	0.706784181	0.552949651	-37.64858915	52.44858915	-37.64858915	52.44858915
X	0.02	0.186547581	0.107211253	0.924407105	-0.782649459	0.822649459	-0.782649459	0.822649459

Figure 8-13. *Output of an example of no correlation using the Pearson correlation method*

In Figure 8-14, the multiple R is 0.075 out of 1 in and is close to zero. The zero direction is shown in the graph, as well as that the Y variable goes almost straight across the x-axis. The P-value for the Pearon's statistical test results is 0.924, which is greater than the alpha of 0.05. This demonstrates that there is no significant linear relationship between the variables X and Y.

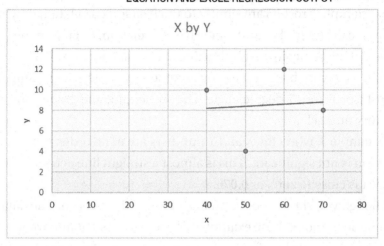

Figure 8-14. *Linear trend of a no correlation using a Pearson correlation method*

Step 4: Results

The first two Pearson correlations are significant (r = 0.9867 and r = -0.9827) and are very close to 1 and negative 1. It is the same correlation, only one is a positive and the other is a negative correlation. The third is not (r = 0.0756) and is very small and close to zero; it displays on the graph as almost horizontal or zero slope. This means that there is no linear relationship, which is demonstrated by a P-value of 0.924, which is a number that is larger than alpha of 0.05.

Step 5: Descriptive Analysis

By looking at the data and the slope, you can see from example 1 that while the X variable is decreasing, the Y variable is increasing for the negative correlation. This is almost a negative 45-degree angle. The scatter plot and graph show the same story more clearly.

The example 2 results are similar to the Pearson correlation or what the output calls multiple r as the negative correlation, but it is in the positive direction. Again, just by looking at the data and the slope, you can see while the X variable is increasing and the Y variable is increasing. This is almost a positive 45-degree angle. The scatter plot and graph show the same story more clearly.

In example 3, when the X and Y numbers are in no order, the correlation is not significant. This is almost a straight line across. And its multiple r is close to zero, or 0.076.

Although you should not be analyzing small datasets and the datasets should be 30 or more, these examples do give you insight into what happens when the order of the data changes by just a bit. To be more exact, there are statistical techniques to use that will enable the researcher to determine how large of a sample to use. They are not reviewed in this book.

Summary

The Pearson correlation is the linear association between the independent variable and the dependent variable. The Pearson correlation has a sign of positive, zero, or negative. Its strength is between +1 and -1. It also has a linear trend. Although the demonstrations used here were from very small datasets, the data from the independent and dependent variables should each be from a dataset that is 30 or more. The last three examples were for demonstration purposes only because they are from very small datasets. As with the example presented of the items bought and total revenue, the relationship between the two variables must be of some importance, must be meaningful, and must help solve a business question.

CHAPTER 9

Independent T-Test

Chapter 7 introduced the idea of thinking through a business question and building a hypothesis test around that query, and it elucidated on the five steps to run a statistical inquiry. This chapter shows how to implement those five steps using the statistical test referred to as the *independent t-test* or the *student's t-test*. Again, a small dataset is used for demonstration purposes only to show how to use the independent t-test and understand its purpose. You'll want to use a larger dataset in real-life situations.

Independent T-Test at a Glance

An independent t-test looks at the difference of two independent sample means over its standard error. *Independent* means that the two samples are not dependent or related, and the samples are from different groups and do not overlap. Again, the data needs to be from a random sample. Moreover, as the sample size increases, the t-distribution starts looking more and more like the normal distribution. An independent t-test is used when there is a dichotomous variable and a continuous variable (refer to Appendix A), and the difference between the independent means of the dichotomous variable or the two groups are determined to be significantly different (if the alternative hypothesis is true) or there is not enough evidence to determine if the two means are different (if the null hypothesis holds). A few concepts to know about the t-distribution are that the mean is zero, just as the mean of the standard normal distribution is zero and the distribution is symmetric about the mean as is the normal distribution.

© Rhoda Okunev 2022
R. Okunev, *Analytics for Retail*, https://doi.org/10.1007/978-1-4842-7830-7_9

Furthermore, as the size of the sample increases, the standard deviation approaches closer and closer to 1 as the standard normal distribution is 1. This is because as the sample increases, the t-distribution looks more and more like the standard normal curve until it becomes large enough to approach the normal curve distribution.

The t-distribution, like the standard normal, goes from negative infinity to positive infinity and never hits the x-axis. Lastly, the t-distribution appears like the normal curve with a mirror image on both sides, except it is smaller and the sample is less peaked in the center (which is the mean) than the normal distribution, and it has fatter tails than in the standard normal distribution. As the sample size increases, the curve for the t-distribution changes and starts looking more and more like the normal curve.

The independent t-test is used when the small sample sizes of the two groups are 30 or more when it is known that the distribution is from a normally distributed distribution or both. Again, as with the Pearson's r correlation, there are more rigorous techniques for determining the sample size that are not reviewed in this book. Again, you can "eyeball" a histogram to see if the distribution looks approximately normally distributed. A statistician will have other methods that they can use that are not discussed in this book.

Hypothesis Test

This is an example of a hypothesis test of the difference between two independent means of revenue of stores and online sales channels. It consists of the five steps previously introduced in Chapter 7, all of which are covered in the following section.

Step 1: The Hypothesis, or the Reason for the Business Question

The test shows that the independent means from the two channels are either H_0 (not enough evidence to say the two independent group means are different) or H_1 (two independent group means are significantly different from each other). Here again a two-tailed test is used because there is the equal sign for the null hypothesis and a not equal sign for the alternative.

$$H_0: \mu_1 = \mu_2$$
$$H_1: \mu_1 \neq \mu_2$$

Step 2: Confidence Level

Just like the Pearson correlation has degrees of freedom and confidence level, so does the independent t-test. For the t-test, the standard alpha of 0.05, which is a 5 percent significance level, is used as the standard on any statistical package. Since there are two groups with four subjects in each group, both groups are added together with two degrees of freedom subtracted (one for each group). The degree of freedom is used because of the use of a sample, and, therefore, constraints are put on sample statistics.

$$df = n_1 + n_2 - 2 = 4 + 4 - 2 = 6$$

Some computer programs use a more mathematical formula to calculate the degrees of freedom, and the result is a bit different. This method is used when there is not a computer that calculates the exact degrees of freedom. The standard two-tailed test is used in this example.

Step 3: Mathematical Operations and Statistical Formula

The main objective for the independent t-test is to determine the difference between the two independent group means over the standard error, which is the noise in the formula. The first step is to calculate the independent mean for each group. The next step will be to calculate the standard error.

Here, there is the variable called "shop" that shows the difference between the binary mode of shopping: Internet versus store shoppers. The object is to determine from the data given if there is a significant difference between Internet and store shoppers in terms of how much they spend or their revenue.

The first part shows the data. The second part introduces the formula for the independent t-test. Then the means are determined, and the step-by-step procedure to figure out the variance and standard deviation for each group is followed so that the pooled standard error can be calculated. Once that is done, step 4 of the results can be analyzed.

1. Table 9-1 shows the data.

Table 9-1. *Data for Type of Shop and Total Revenue*

Shop	Total revenue		
1	90	$1 = X_1$	= internet
1	100	$2 = X_2$	= store
1	110		
1	95		
2	70		
2	50		
2	45		
2	65		

2. The formula is shown here.

The X-bar is the mean of each sample, while S^2 is the variance for each sample, and S is the standard deviation.

$$t = \frac{\text{X-bar}_1 - \text{X-bar}_2}{\text{Sqrt}\{[((n_1 - 1) S_1^2 + (n_2 - 1) S_2^2)/(n_1 + n_2 -1)](1/n_1+n_2)\}}$$

$$\text{X-bar}_1= \frac{90+100+110+95}{4} = 98.75$$

$$\text{X-bar}_2= \frac{55 + 70 + 65 + 45}{4} = 58.75$$

S is the standard deviation:

$S = \text{sqrt}[(x – \text{x-bar})^2]/(n–1)]$

S^2 is the variance:

$S^2 = (x – \text{x-bar})^2]/(n–1)$

Table 9-2 shows how the standard deviation is calculated for the group store and Internet group.

Table 9-2. *Independent T-Test Calculated Using Excel Spreadsheet*

Internet	X1	X1bar	X1-X1bar	(X1-X1bar)²	Store	X2	X2bar	X2-X2bar	(X2-X2bar)²
1	90	98.75	-8.75	76.5625	2	55	58.75	-3.75	14.0625
1	100	98.75	1.25	1.525	2	70	58.75	11.25	126.5625
1	110	98.75	11.25	126.5625	2	65	58.75	6.25	39.0625
1	95	98.75	-3.75	14.0625	2	45	58.75	-13.75	189.0625
		Sum	0	218.75			Sum	0	368.75

3. Here the variance is calculated for each group:

Group 1 variance: $S^2 = 218.75/3 = 72.92$, $S = sqrt(218.75/3) = 8.54$
Group 2 variance: $S^2 = 368.75/3 = 122.92$; $sqrt(368.75/3) = 11.087$

4. When the variances are not considered different,
    the pooled standard error formula is calculated.
    However, when the variances are considered
    significantly different, there is a more complicated
    formula that needs to be computed. The book
    does not go over the math when the variances are
    different, but at times this independent t-test with
    different variances needs to be used because the
    variances are statistically different. (Note: When
    the variances are considered significantly different,
    the Data Analysis ToolPak can do that computation
    as well.)

The following are the calculations for the pooled standard error:
$Sqrt\{[((n_1 - 1) S_1^2 + (n_2 - 1) S_2^2)/(n_1 + n_2 -1)](1/n_1+n_2)\}$
$=sqrt\{[3(218.75/3)+ 3(368.75/3)]/(4+4-2)*(1/4 + 1/4)\}$
$=sqrt\{ [218.75 + 368.75]/6]*0.05\}$
$=sqrt\{587.5/6*(0.50)\}$
$= 6.9970$
Independent t-test = 98.75 - 58.75   40

$$\frac{}{6.997} = \frac{40}{6.997} = 5.72$$

In Table 9-3 you can see the output and results of when a t-test is
conducted with the Data Analysis ToolPak on the Data tab.

Table 9-3. *Output of the Independent T-Test (Using ToolPak)*

t-Test: Two-Sample Assuming Equal Variances

	X	Y
Mean	98.75	58.75
Variance	72.91667	122.9166667
Observations	4	4
Pooled Variance	97.91667	
Hypothesized Mean Difference	0	
Df	6	
t Stat	5.716717	
P(T<=t) one-tail	0.000621	
t Critical one-tail	1.94318	
P(T<=t) two-tail	0.001241	
t Critical two-tail	2.446912	

The independent t-test result = 5.72.

The level of significance for the t-test = 2.447.

The p-value = 0.0012141.

Step 4: Results

The results are significantly different at the 0.05 alpha level since the p-value is less than the alpha level. Another way to say this is that the two groups are significantly different because the alpha is 0.05 and the p-value is 0.0012, and 0.0012 is smaller than 0.05; therefore, the two groups are significantly different.

Step 5: Descriptive Analysis

This result shows that shoppers spend significantly more on the Internet (X-bar$_1$ = 98.75) than in the stores (X-bar$_2$ = 58.75). A bar graph could have been used here to show the difference in the means. These results are straightforward and expected. Therefore, the company may want to have more and extra merchandise for their online shop than in their retail stores. This may help the company cut costs because a retail store is more expensive to upkeep than an online store.

Summary

In the next chapter, there will be a more involved example that will put all these statistical tests to use and show how to use these tests in a real life-life scenario. The next chapter will show a business situation of a retail email campaign.

CHAPTER 10

Putting It All Together: An Email Campaign

The objective of this example email campaign is to see whether the buying habits of customers are affected by the type of model used in advertising, with an A/B-style test format. The campaign randomly splits up the creative content delivered in an email campaign. One version of the email shows pictures of models wearing a dress. The other version of the email does not have models; it just shows the same dress on a mannequin.

Examples of the email campaign (in the form of an Excel spreadsheet) can be found in Appendix B at https://github.com/Apress/analytics-for-retail. You need to download this spreadsheet to follow along in the chapter.

Test Goal

The study uses A/B testing to see whether the merchandising in the email impacts click-through, conversion rate, orders, items sold, and revenue, although other factors could affect the outcomes. A/B testing is a form of randomized controlled study, and it is often used in retail marketing.

© Rhoda Okunev 2022
R. Okunev, *Analytics for Retail*, https://doi.org/10.1007/978-1-4842-7830-7_10

This A/B testing is when there are two types of users assigned randomly, in this case the model and no models, on the landing page to determine the business questions of which method is more successful in attracting buyers to purchase items and increase revenue.

The business questions of this study uses the follows analysis: When using A/B testing, which prototype (model versus no model) affects the click rate, conversion rate, orders purchased, dresses sold, and revenue generated? Also, is there a correlation in the positive direction between conversion rate and revenue for the model and the no model group?

Method

Because these customers are early-morning shoppers, emails were delivered for the month in the early morning at the same time and day of week to both groups as a control. Both groups had the same email subject line and landing page and could not see whether there was a model until they opened the email campaign. Here the campaign is trying to control the variables and show both groups the same content, except for the presence or absence of a model, to determine if models really contribute to customers' buying habits.

Key performance indicators measured for the campaign include deliverability rate, open rate, click-through rate, conversion rate, and revenue generated. This is achieved via embedded tracking pixels from your CRM system onto the website to attribute on-site behavior and revenue to the campaign.

Since randomized A/B testing was used for both the model and no model image group, the click-through rate to the same website landing page from the email for the average time on-site was controlled.

Data Constants

This type of CRM data can be downloaded to Excel and uploaded from Excel to/from the CRM system, alongside Google Analytics or Core Metrics to run the following statistics.

The dataset for the email campaign presented here was checked, and each group has 30 data points. There were no missing values. There were no outliers or extreme values in the dataset. The minimum, maximum, and range variables for conversions (minimum=5.1, maximum=6, range=0.9), orders (minimum=25, maximum=27, range=2), dresses sold (minimum=30, maximum=41, range=11), and revenue (minimum=3000, maximum=4100, range=1100) for the model, as well as the variables for conversions (minimum=3.6, maximum=6.0, range=2.4), orders (minimum=21, maximum=23, range=2), dresses sold (minimum=22, maximum=30, range=8), and revenue (minimum=2200, maximum=3000, range=800) for the no model group, all looked in the normal scale for each category. The data appeared approximately normally distributed on the histogram (the charts are not shown in the book), and the means and medians were similar for the variables for conversions (mean=5.56, median=5.55), orders (mean=26, median=26), dresses sold (mean=36, median=36), and revenue (mean=3583, median=3550) for the model group as for the variables for conversions (mean=4.8, median=4.7), orders (mean=22, median=22), dresses sold (mean=26, median=26) and revenue (mean=2603, median=2600) for the no model group. The normal curve is a special case where the mean, median, and mode are the same numbers or in real-life approximately the same. (The mode was not calculated because Excel's mode function does not always work.) Refer to Appendix B for the dataset and the descriptive statistics for the email campaign. At this point, the statistical analysis techniques are ready to be used to describe and analyze the data, the results will be summarized, and the findings will be explained for each business question. The statistical output is in Appendix B for each procedure.

Type of Shopper Targeting

From the store's database of 300,000 women at a midsize store, the population of loyal shoppers between the ages of 25–35 is 120,000 customers. A previous segmentation study using shopping history derived from loyalty cards had shown that this age group would be most interested in this type of dress and price point from previous buying habits. (Separate campaigns may be sent out to other targeted groups.) This cohort is randomly split into two groups.

Many stores use the entire population of customers, not a targeted segment, to perform this operation. This is because they want the entire population to receive the sales messaging.

Time of Year and Duration

The emails were sent daily throughout the month of November when shoppers begin shopping for the upcoming holiday season.

Cost of Dresses

This site sells only dresses, and there were 2,000 dresses at the beginning of the campaign. All dresses at checkout were adjusted to $100. No promotional code was necessary.

The wholesaler's original cost to the store was $69.44 for each dress. The store marked up the dress 80 percent to $125. For this promotion, dresses are discounted 20 percent to $100 each.

Medium Type

Shoppers can buy these dresses only on the Internet.

Steps to Assess the Success of the Email Campaign

Both the no model and the model key performance indicators were included. These key indicators involved who was the targeted group, how many emails were sent, when the customer would receive the email, what percentage received and opened their emails, and the conversion rate calculated.

Here are the no model metrics on the key performance indicator analysis:

- **List size:** A random sample of 15,000 women per day with a history of buying dresses at the company's stores is used. The same women are targeted every day.

- **Delivered rate:** Approximately 90 percent, or 13,500 women, received the email per day.

- **Open rate and kept subscription to emails:** Of the 13,500 women who received the email, 15 percent (2,025 women) opened their email per day.

- **Click rate:** Of the 2,025 women who opened the email, 23 percent (461 women) clicked through to the dresses per day.

- **Conversion rate:** Conversion = (Number of orders)/ Sessions*100 /day = approximately 22/461*100=4.77%/ day. The session is over once the browser is closed. Calculate the conversion rate per day. After 30 days, all orders were processed to determine the campaign's effect.

Here are the model metrics on the key performance indicator analysis:

- **List size:** A random sample of 15,000 women per day with a history of buying dresses at the company's stores. The same women are targeted every day.

- **Delivered rate:** Again, approximately 90 percent, or 13,500 women, received the email.

- **Open rate and kept subscription to emails:** Of the 13,500 women who received the email, 13 percent (1,755 women) opened the email per day.

- **Click rate:** Of the 1,755 women who opened the email, 27 percent (468 women) clicked through to dresses per day.

- **Conversion rate:** Conversion = (Number of orders)/ Sessions * 100/day = approximately 26/468*100=5.56%/ day. The session is over once the browser is closed. Calculate the conversion rate per day. After 30 days, all orders were processed to determine the campaign's effect.

Statistics Conducted: Results and Explanations

An independent t-test was conducted here to determine whether there is a difference between the means of the two groups' conversion rate, revenue, dresses sold, and ordered dresses. A two-tailed test was used for all the t-tests. A Pearson correlation test was conducted for conversion rate and revenue to see if there was a relationship.

Independent T-Test 1: Conversion Rate Between Models and No Models

An independent t-test was conducted here to determine if there is a difference between the mean conversion rate of the two groups.

- **Step 1: The Hypothesis, or the Reason for the Business Question:**

  - **Null hypothesis:** $\mu_1 = \mu_2$

  - Meaning: There is not enough evidence to determine if there is a difference in the means of the conversion rate between the model and no model conditions.

  - **Alternative hypothesis:** $\mu_1 \neq \mu_2$

  - Meaning: There is a significant difference in the mean conversion rate between the model and no model versions.

- **Step 2: Confidence Level:** Alpha = 0.05

- **Step 3:** See "Mathematical Operations and Statistical Formula" in Appendix B.

- **Step 4: Results**: There was a significant difference between the model/no model's mean rate conversion rate.

- **Step 5: Descriptive Analysis:** There was a significant difference between the mean conversion rate of the model (5.56) and the no model (4.77). There was a higher conversion rate for the models. In other words, more shoppers bought when there was a model. The shoppers saw a picture and liked how it looked on a real human.

Independent T-Test 2: Revenue Between Models and No Models

An independent t-test was conducted here to determine if there is a difference between the mean revenue of the two groups.

- **Step 1: The Hypothesis, or the Reason for the Business Question:**

  - **Null hypothesis: $\mu_1 = \mu_2$**

  - Meaning: There is not enough evidence to determine if there is a difference in the means of the revenue between the model/no model conditions.

    - **Alternative hypothesis: $\mu_1 \neq \mu_2$**

    - Meaning: There is a significant difference for the mean revenue on-site between the ones that used models or no models.

- **Step 2: Confidence Level:** Alpha = 0.05

- **Step 3:** See "Mathematical Operations and Statistical Formula" in Appendix B.

- **Step 4: Results:** There was a significant difference between the revenue of the model and no model groups.

- **Step 5: Descriptive Analysis:** The model group had a higher mean (3,583) than the no model (2,603) group, and there was a significant difference between the two means. This shows that when you use a model, there are better financial gains than when a model is not used.

Independent T-Test 3: Dresses Sold Between Models and No Models

An independent t-test was conducted here to determine if there is a difference between the mean dresses sold of the two groups.

- **Step 1: The Hypothesis, or the Reason for the Business Question**

  - **Null hypothesis:** $\mu_1 = \mu_2$

  - Meaning: There is not enough evidence to determine if there is a difference in the means of dresses sold on-site between the campaign that used models and the one that used no model.

  - **Alternative hypothesis:** $\mu_1 \neq \mu_2$

  - Meaning: There is a significant difference for the mean dresses sold on-site between the campaign that used models and the one that used no model.

- **Step 2: Confidence Level:** Alpha = 0.05.

- **Step 3:** See "Mathematical Operations and Statistical Formula" in Appendix B.

- **Step 4: Results:** There was a significant difference between the models and no models on mean dresses sold.

- **Step 5: Descriptive Analysis:** The groups had a significant difference in their mean dresses sold between the model (36) and no model (26). Therefore, the presence of a model or no model did positively affect the average number of dresses sold.

Independent T-Test 4: Orders of Dresses Between Models and No Models

An independent t-test was conducted to determine if there is a difference between the mean number of dresses ordered between the two groups.

- **Step 1: The Hypothesis, or the Reason for the Business Question:**

  - **Null hypothesis: $\mu_1 = \mu_2$**

  - Meaning: There is not enough evidence to determine if there is a difference in the mean number of dresses ordered between the campaign that used models and the one that used no model.

  - **Alternative hypothesis: $\mu_1 \neq \mu_2$**

  - Meaning: There is a significant difference in the mean number of dresses ordered between the campaign that used models and the one that used no model.

- **Step 2: Confidence Level:** Alpha = 0.05.

- **Step 3:** See "Mathematical Operations and Statistical Formula" in Appendix B.

- **Step 4: Results:** There was a significant difference between the number of dresses ordered between the models and no models versions.

- **Step 5: Descriptive Analysis:** The groups had a significant difference in their mean number of dresses ordered between the model (26) and no model (22). Therefore, the presence of a model encouraged the buyer to buy more dresses per order.

Pearson Correlation by Model: Relationship Between Conversion Rate and Revenue

A Pearson correlation was performed to determine if there was a significant relationship between conversion rate and revenue on the site.

- **Step 1: The Hypothesis, or the Reason for the Business Question:**

  - **Null hypothesis: p = 0**

  - Meaning: There is not enough evidence to assume that there is a relationship between conversion rate and revenue.

  - **Alternative hypothesis: p ≠ 0**

  - Meaning: There is a significant correlation between conversion rate and revenue.

- **Step 2: Confidence Level:** Alpha = 0.05.

- **Step 3:** See "Mathematical Operations and Statistical Formula" in Appendix B.

- **Step 4: Results:** There was a significant positive relationship between conversion rate and revenue when a Pearson correlation was conducted separately for both the models (0.75) and no models (0.62).

- **Step 5: Descriptive Analysis:** This shows that whether a model was present or not, as the conversion rate increased, so did the revenue. This is what the analyst would hope to find.

Notice that the scale in Figure 10-1 is different from the scale in Figure 10-2. This is because the graph would not show its positive increase as clearly if the graphs' scales were the same. But both charts show a positive increase in the linear trend.

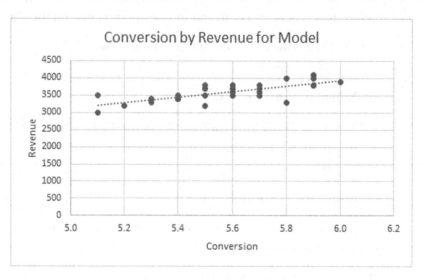

Figure 10-1. *Conversion by revenue for model*

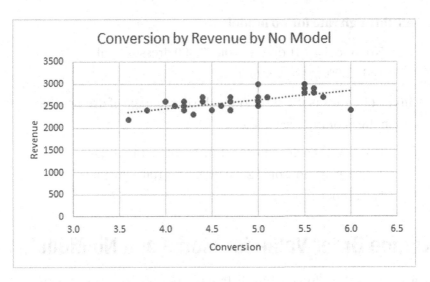

Figure 10-2. Conversion by revenue for no model

Sell-Through Rate for Model and No Model

The formula for sell-through rate gauges how fast the company sells its products. It aids the company to predict its demand in that product. Sell-through rate measures the percentage of inventory that is sold compared to the inventory that is sent from the manufacturer. Most companies look very carefully at this percentage. This percentage lets the company know how many products to purchase or that there is a problem selling certain products (although it does not indicate what the issue is).

Sell-through rate = units sold/(units sold and units on hand) × 100

Sell-through rate for the model:

> Model = 1,077 dresses sold/2,000 dresses total × 100
>
> = 54% for November

Sell-through rate for no model:

> No model = 781 dresses sold/2,000 dresses total ×
> 100 = 39% for November

This last figure lets the store know the percentage of dresses that were not sold in the ecommerce business.

> Dresses not sold = 2,000 – 1,077 – 781 = 142. Take
> 142 dresses not sold/ 2000 dresses × 100 = 7% for
> November.

Average Order Value for Model and No Model

The average order value is a metric that tracks every dollar spent by customers with every purchase the customer makes. This is a benchmark, and it indicates that as the average inventory value increases so does the profits for the company.

AOV= Average purchase = total revenue/ total orders

Average order value model:

> Model = $107,500/780 = $137.82 or $138 for
> November

Average order value no model:

> No Model = $78,100/661 = $118.15 or $118 for
> November

The average order value for the customer is higher for the model than the no model, which is to be expected because with the model the customer has an opportunity to see the garment worn on a model and has a better idea how the garment would look on them.

Note: To have a large enough sample, it is typically recommended that 30 or more data points be used for each type of model group. Depending on the size of your database, more sophisticated sample size calculations can be considered.

Total Metrics on Key Performance Indicators for Email Campaign

The key performance indicators (KPIs) are important vital factors that allow a company to improve their strategies and operations.

Conversion rate is a major key performance indicator, and it shows how many of your site visitors or campaign recipients actually make a purchase (Table 10-1).

Table 10-1. Basic Metrics for Email Campaign

Mean Type of Model	Conversion Rate per Day	Average Click Rate
1 (no model)	4.77%	23%
2 (model)	5.56%	27%

The conversion rate is the number of conversions divided by the number of visits multiplied by 100. The no model conversion rate would be approximately 0.0477/day, or approximately 4.77 percent/day, and for the model the conversion rate would be approximately 0.0556/day, or approximately 5.56 percent/day. November has 30 days. Therefore, 30 conversion rates and 30 response times for each type of model are needed.

Average Click-Through Rate

The average click rate is the average number of times customers will click on the email link to the total number of customers who received the email for the email campaign.

Type of Model

The no model's click-rate range is larger, and the standard deviation or variability is larger than with the model. This is expected because for the model version because the customer has an idea of what the product is they are clicking on to accrue.

- 1 = no model (range of 40; it is between 100 and 140 avg click rate and has a bigger standard deviation).

- 2 = model (range of 10; it is between 115 and 125 avg click rate and has a smaller standard deviation).

Profit per Dress

The profit per dress is the sale price the store is selling the item minus the wholesale price it cost to buy the item from the wholesaler. The more the company can sell the dress for and the less the original cost, the better for the retailer (Table 10-2).

Table 10-2. *Revenue or Profit from Campaign*

Dresses sold times the cost of the dress	
Model	**No Model**
30.56 dress cost × 1,077 dresses sold	30.56 dress cost × 781 dresses sold
= $32,913.12	= $23,867.36

Profit per dress = sales price of dress-original wholesale = $100 – $ 69.44 = $30.56

Total profit from campaign = $56,780.48 = $32,913.12 + $23,867.36

ROI and ROAS

Return on investment (ROI) and return on ad spending (ROAS) are two essential metrics for measuring the results and profits for the campaign that is run. ROI is the net profit divided by the net spending multiplied by 100. ROI includes the cost of software, designs, and people.

ROAS is revenue made from ads divided by advertising spending. ROAS does not tell you if the paid advertising is effective, but it explains if the ads are effective in driving clicks and revenue.

The higher the ratios for both the ROI and the ROAS, the better. This means the company is gaining a higher percentage in revenue than cost for the ROAS and also bringing in more profit than investment for the ROI.

Here are the ROAS and ROI formulas for the model and the no model:

ROAS = revenue derived from ad campaign/cost of ad campaign

ROI = profit derived from ad campaign/total cost of investment from campaign

	Model	**No Model**
ROAS	= 32,913/6,570.4	= 23,867.36/5,966.84
Ratio	5:1	4:1

(The example reads as follows for the model: for every $1 spent on the ad campaign, $5 is gained in revenue.)

ROI	= 32,913/8,213	= 23,867.36/7,955.79
Ratio	4:1	3:1

The ratios for the model are higher than the ratios for the no model version.

Summary and Discussion on Results

We had a business question of, does showing the model encourage the buyer to purchase more? Using the A/B testing method, we showed that it does. This was demonstrated because the model group had statistically significant different means that were larger for conversion rate, orders made, dresses sold, and revenue acquired than the no model group.

The other business question was, does the conversion result in more revenue? This showed a significantly positive association for both the model and no model groups, which is also a good sign. This means that as conversion rate increased, so did revenue in the positive direction. If there was a negative correlation this would mean that discounts were driving the increase of conversion rates. There were not many dresses left after the campaign, and the sell-through rate for the model was better than the no model version. The model performed better than the no model version for the sell-through rate, AOV, average click rate, and profit. The ROI and ROAS were better for the model version than the no model version. These findings are to be expected because the buyer was given clearer information about what they were purchasing.

Thoughts for Further Analyses

This email campaign was successful and, for the most part, came out as expected. As anticipated, the model group mean revenue, conversion rate, orders, and dresses sold were all more than the no model version. And for both the model and no model versions, there was a positive relationship between revenue and conversion rate. This means the more interesting the merchandising and more information the buyer has, the more they could picture the dresses on themselves and decide to buy them.

Moreover, the brand values that the store is selling are key to understanding and analyzing the results of the email campaign. It is imperative to keep the brand desirable and relevant or, at least, consistent and sustainable in order to retain and continue to increase the number of loyal customers. The object is to sell products and increase revenue while controlling or minimizing costs and returns, while also reinforcing brand value such as best price, best quality, or some other core brand feature.

Summary

The focus of the email program needs to be on retaining customers in the database and on growing the database with potential and new customers. Emails tests such as the one shown in this chapter help to optimize merchandising to maximize conversion rate and sales from each campaign. To increase direct sales at the program level, the store manager could use email sign-up on the website to retain new visitors and optimize copy with search keywords to help attract potential customers.

The astute manager will need to run year-over-year predictive analysis, using variance and percent variance to make sure the company is maintaining and increasing its revenue based on the metrics of the key performance indicators, which will be reviewed in the next chapter. Month-over-month analysis may be conducted as well, but keep in mind that sales are different at various times of the year, and each month has a different number of days. The next chapter will cover how to create scenarios for anticipating sales and forecasting into the future.

CHAPTER 11

Forecasting: Planning for Future Scenarios

In this chapter, a statistical forecasting methodology is presented that utilizes past sales in order to predict future sales when the past trend tends to continue normally and there are no black swans, which could be natural disasters, extreme weather, and other events beyond the entrepreneur's control.

The following methodology details how to calculate a twelve-month moving average and then how to smooth the data for a seasonal forecasting approach of sales for around six months into the future by using regression statistics to predict future sales. The forecasting analysis should utilize at least three years of monthly sales data (if you are a new company), but five or more years of monthly sales history is even better to have so that repeating trends can emerge, particularly if the company was recently started. After the forecast method is presented, the chapter covers a few techniques as to how to estimate sales for certain types of scenarios, campaigns, and customers.

This chapter is a companion to the forecasting spreadsheet (in the form of an Excel spreadsheet) in Appendix C at `https://github.com/Apress/analytics-for-retail`. It should be read side-by-side with the full data illustration.

© Rhoda Okunev 2022
R. Okunev, *Analytics for Retail*, https://doi.org/10.1007/978-1-4842-7830-7_11

Regression at a Glance

Regression analysis is a basic statistical tool that helps predict expected behavior based on prior linear trends. The smoothing technique is used to remove random variation in the data in a way that allows trends and cyclical trends to show. Real-life data contains anomalies and extreme events that impact performance both positively, e.g., a one-time government stimulus check that provides extra liquid income for consumers, or negatively, such as in the case of a natural disaster or economic instability. Part of your data collection process will be to try to understand which types of extreme events may occur and need to be understood further. Of course, many extreme events may not be able to be predicted in advance. The most important thing that you should look for as an entrepreneur or manager is an opportunity to impact growth in a positive way, structurally and for the long term. These are events that will shift your regression lines steadily upward instead of providing one-time spikes.

Establishing Data Collection

In the beginning, when organizing a company, it is imperative to set up data collection so that the entrepreneur will be able to undertake all analysis to effectively predict scenarios into the future to sell products and ultimately identify new customers to buy the product. Always think about what sets your company apart from other similar companies and determine how your company is able to differentiate itself from the others so that you know the "magic" that is increasing growth sales. This is key to your brand's value so you can differentiate your brand in the marketplace and the consumer's mind. It will also be helpful as you compare your data to that of your competitors or other benchmarks and can help you understand your performance in context. Benchmarks can include market performance based on markets such as S&P500, Nasdaq, Dow Jones, and other publicly available data such as inflation indexes and other market analysis tools.

Most companies try to predict sales performance just using ideas in their head of how much product they will need by changing various product percentages around and estimating what they think would be the best number of items to have in stock to increase sales due to last year's numbers and holidays and other situations that are about to take place. Those companies just adjust numbers depending on whether they are selling more or less of products and based on which products are the prime sellers. Although this method may be typical, it is not a scientific regression model.

Many companies, even large public companies, do not for the most part use scientific approaches to predict sales because doing so can be costly to update software and can be complicated; a data scientist may need to be hired, and it may be time-consuming to change long-standing processes in operations. Therefore, most often the brute-force method—manual review of sales performance and estimation of future performance based on arbitrary performance targets—takes precedence. In the following section, we'll review a scientific regression approach that deseasonalizes the raw sales data and then uses a time-series moving average method to smooth the data. After that, the regression model is used, and then the data is re-seasonalized with the adjusted-averaged ratio. At first, the reader may find this complicated and cumbersome. This chapter will show you how to do it in a step-by-step process.

Predictive Analysis Using the Spreadsheet

Columns A through C in the Forecasting tab of Appendix C illustrate the time period since the inception, which is the time period that is used to predict the associated sales for that time (column D). Column E is a formula for calculating the moving average and smoothing the data. Since there are 12 months and the aim is to look at the data month over month, a

12-month moving average is used. The first six and last six month's cells are to be left blank in column E. Starting in the seventh cell, the formula =AVERAGE(AVERAGE(D5:D16), AVERAGE(D6:D17)).

Column F calculates the ratio of the sales recorded divided by the smoothing sales prices just calculated. Next, in column G, the unadjusted indicator is calculated by taking all of one particular month at a time and then averaging all of the same month's ratio numbers as possible. The book uses the averageIF function of AVERAGEIF(range, criteria, [average_range]) to do this, but a lookup table method will work as well. For example, for all the January months (month 1), a ratio would be created, and the ratio would include only the months of January. The next would be to take all of the February ratios, etc. Some months may have more years to calculate the average than others, and that is fine.

Once that is done, we need to adjust the numbers to equal 12 because there are 12 months. This is done by summing up the unadjusted number (in cell G2) by the first 12 months and then dividing that number by 12. (The data does not have to start in the month of January like this example.) Then for column Adjusted (H4), take the unadjusted number (G4) and adjust it by multiplying it by the unadjusted number (G2) over the adjusted number (H2).

The next step is to compute the deseasonalized sales by dividing the original sales (column D) by the adjusted indicator (column H).

The next part is a little tricky and involves a little knowledge about linear equations. You need to use the Data Analysis ToolPak in Excel. This technique involves an understanding of the regression function used in the Pearson regression chapter eight and the linear regression equation derived here. The Data Analysis icon is on the Data tab in Excel. Once the Data Analysis ToolPak is downloaded, click it and find the spreadsheet's DRegression tab. Since the linear equation is Y-hat = m*x + b, the b is the intercept of the deseasonalized sales, the m is the slope that Excel calculates for the user, and the x is the time variable. On the

DRegression tab, to perform the regression use for the Y-hat variable, the deseasonalized sales coefficient ($I$4:$I$64), and for the x variable, use the period data coefficient ($A$4:$A$64). Run the regression to output on a separate page and on the bottom will be the coefficients of the b-intercept and the slope called *period*. Put all the numbers into the equation, and for this data set it will be Y-hat = $1,347x + $41,616. The small x is each individual x from line 1 to line 64 and whatever may be used to estimate the linear equation and the future. For instance, when x = 1, the linear trend line will be $42,963=$1,347(1)+$41,616.

To analyze the future (column K), multiply the linear trend line with the adjusted ratio. Copy down the adjusted ratio to the predicted column, and copy down the x's (in column 1) to the number of columns to predict. There are six months projected on this spreadsheet. To see the trend a bit more clearly, in column L and column M, the book calculated the previous month's rate of change and the rate of change year over year. Now you have the predicted seasonalized data. It is always a good idea to graph the data to make sure its trend is linear before you put back the seasonality of the data. Once the seasonality is reloaded, it is important to notice that it follows the trend of the original sales but is smoothed out. If the company is doing well with an upward trend as this company is doing, an upward consistent trend will appear. The data can be shown as an estimate for how it is believed that the company should continue its trend if the market forces remain relatively consistent and normal.

Scenario Analysis

Most firms at this point will conduct some form of stress testing of the scenarios, campaigns, and types of customer analysis, as shown on the next Excel tabs. As is evident with the Scenarios tab, the percent growth per month is the same percent as the variance on the bottom of the page. Since the percent increase on the Forecasting tab is for an average increase around 15 percent for many months, the Scenarios tab uses 10 percent as a lowball (this percent can easily be changed), the 15 percent as normal, and then some reasonable scenarios the company would like to have. Those percentages can easily be changed by typing in the green box the percentage to assess what the company believes it should assess. The boxes in green are the numbers and percentages, and they can be changed.

Campaign Analysis and Prediction

The Campaigns tab is to determine the potential performance of a month-long campaign broken out by week based on performance targets and a promotional offer. Three numbers can be changed on this spreadsheet. First you need to put in the green B4 cell the estimated sales for January 2xx1 ($150,000) and then put the percent growth target in cell C1 to increase it for January 2xx2 (15 percent) and then put in the actual sales (E6-E9). Columns B and C illustrate the "brute-force" forecast based on the growth target. Column E shows the actual sales performance against the original estimate, and column F uses predicted sales from the Forecasting tab. The actual sales amount can be changed, but remember total customer sales should equal customer sales from the forecasting sheet.

Then the variance in columns H through L shows the gaps in these estimates. Whereas brute-force estimates are used to plan inventory to meet the most optimistic view of the growth target, the variance between the estimate and the data-driven prediction illustrates either performance successes when positive, or the need to readdress strategy when negative. Negative variance to performance estimates could be addressed with strategic adjustments such as increased marketing spending or inventory realignment by category investment.

The total sales for actual sales (cell E4) is read from the Forecasting tab. The variance table compares the differences between the estimated numbers and the actual numbers. The percent variance is the rate of change of the estimated numbers and the actual numbers. While each month has an overarching target, the brute-force estimate is often evenly distributed on a weekly basis unless extreme promotional activity is planned. However, in reality, variances depend largely on marketing priorities and exposure, campaign releases, and market demand based on externalities such as weather and inventory availability. This is illustrated in the weekly variances in actual sales in column E and used to predict weekly variances for the prediction for the future year. These nuances can help identify opportunities for strategic optimization on a week-by-week basis.

Consumer Analysis and Prediction

The Customer tab was developed based on the 80/20 rule, which suggests that 80 percent of sales will typically be garnered by 20 percent of customers, typically your most loyal. Then there is the group of customers that buy mainly when the products go on sale, and the others are customers who buy once in a while or are new customers who the company would like to convert to loyal customers. While the company should always be thinking of ways to increase their customer base, it is important to try to estimate the sales for each group of customers. The total sales for 2xx1, Base Year 2xx1 (B4), comes from the forecasting spreadsheet. The percentages in dark green and yellow can be changed to fit the store's expectations and size of the customer base. This can also help you plan unique messaging to address each customer group's priorities and affinities for your brand, whether it is based on price and discounting or on quality of service, product, and brand values alignment.

Summary

This was a difficult chapter, and you needed a basic knowledge of regression to get the most out of it. All companies need and want to expand so that they can grow. To do this, companies need to predict from past sales what future sales will be. This chapter gave a more rigorous estimate of what sales may look like given that no out-of-the-ordinary events will occur.

CHAPTER 12

Epilogue

This book showed the step-by-step approach to running many of the fundamental statistics, percentages, and analytics for a retail company. Using a data-driven approach to retail management in a systematic way with analytic and predictive methods will give you a deep and consistent understanding of the performance of your business so that you can optimize it. However, the overarching business questions—mission, vision, and values—must be thought out from the start of the development of the company's products and services to be able to clearly identify what it stands for and how it stands out from the rest of its peers. This is imperative for the business to succeed in achieving its stated mission and goals. Of course, a company will always need to be aware of new trends and its competitors and evolve with new technologies, styles, and skills.

The entrepreneurs and the developers of the business want to sell as many of their products as possible in order to make a huge profit, the bottom line. To sell as many products as possible, it is important to understand how to efficiently organize the data, develop the campaigns to optimize sales and win new customers, and analyze the results to determine what worked and what did not. Using data analysis to produce simple and direct graphics can help you gain important takeaways and develop actionable conclusions easily. Developing a wide-ranging and comprehensive set of data and analysis across every functional area is also important to determine what factors have allowed the company to go in a positive or negative direction, as well as where it is going and wants to go in the future. This includes tracking costs and inputs across

© Rhoda Okunev 2022
R. Okunev, *Analytics for Retail*, https://doi.org/10.1007/978-1-4842-7830-7_12

labor and materials, marketing, and operations alongside sales, margin, and turnover to create a holistic picture of the business performance and return on investment so that you understand the time, effort, and resources necessary to produce your results. In the final analysis, it is all about making the best product, differentiating it, knowing your customer and what is needed to successfully sell without compromising your margin or brand values, and increasing the top line, meaning the sales.

To keep the brand desirable and relevant, the sales representatives are the eyes and ears of the company whether on the phone, in the store, or over an email. The customer service and sales departments should be regularly consulted to see what the customers are saying about the products. Salespeople should be shown how to address potential concerns, competitive threats, and key brand selling points with the customers. A key to working with data is understanding the customer's likes and dislikes, which help to provide qualitative information that often illuminates valuable insights to be gleaned from quantitative analysis. For example, when evaluating returns, understanding the reasons why a customer returns a particular item can often illuminate product issues that need to be resolved, reflected in high returns for one particular item, fabrication, or size. Low sales might be an indicator of a better-priced competitive alternative that your customers know of through shopping portals or Amazon.

The brand value proposition—the reasons why your product is attractive to your consumer—must be understood by your internal team, communicated effectively, and also made relevant to the customer. That customer's demographics and psychographics need to be investigated to determine that the right products are being sold to appropriate groups and that there is no misalignment with the targeted groups. This will help retain customers who are interested in the brand and have a desire to purchase the items, as well as attract similar new customers, which will increase top-line sales.

This book aims to predict what the customer will want in the future based on past history. It is important to use vital metrics and use various testing styles to optimize conversions and revenue. As important, the statistical analysis needs to target the appropriate groups for email campaigns and to use various style testing to determine how to increase revenue. Corporate departments need to develop reasonable benchmark KPIs for the managers to achieve. The managers and salespeople need to be trained to develop honest relationships with their customers; to know their likes, dislikes, and levels of satisfaction; and then to relay that information to the analysis team so that the results of the campaign can be more clearly understood. More intricate and detailed questions can be ascertained from the data when the analyst team understands their customers' preferences, profiles, and other qualitative information. In the end, this will increase the bottom line, increasing efficiencies, scaling the customer base, and increasing profit.

Index

A

A/B testing method, 115, 132
Alternative hypothesis, 78, 79, 86, 107

B

Basic math, 24, 38
Bell-shaped curve, 16

C

Central limit theorem (CLT), 16
Confidence level, 79, 86, 109
Cost of goods sold (COGS), 34, 40, 43, 46

D

Descriptive analysis, 13, 81, 92, 105, 114

E

Earnings before income, tax, depreciation, amortization (EBITDA), 39
Efficiency ratios, 54, 62

Email campaign
    A/B testing, 116
    average click rate, 130
    data constants
        assess steps, 119, 120
        cost of dresses, 118
        CRM system, 117
        shopper targeting, 118
    definition, 115
    examples, 115
    independent t-test, 121–124
    KPIs, 133
    method, 116
    Pearson correlation, 125, 127
    profit per dress, 130
    ROI/ROAS, 131, 132
    sell-through rate, 127–129
Entrepreneurs, 143

F

Financial ratios
    debt/leverage, 58, 59
    definition, 53
    efficiency, 62, 63
    liquidity, 54–57
    probability, 60, 61
    profitability, 54

Forecasting methodology
  campaign analysis/prediction, 140, 141
  consumer analysis/ prediction, 141
  data collection, 136, 137
  definition, 135
  regression, 136
  scenario analysis, 139
  spreadsheet, 135, 137, 139
Frequencies, 66
  Excel, 70, 71
  store, 69

G

Gaussian curve, 16

H

Horizontal bar charts, 73
Horizontal charts, 71
Hypothesis tests, 77

I, J

Independent t-test, 72, 79, 107, 108
  confidence level, 109
  descriptive analysis, 114
  hypothesis test, business question, 109
  mathematical operations, 110–112
  results, 113

K

Key performance indicators (KPIs), 37, 40, 116, 129

L

Law of large numbers (LLN), 16, 19
Least square line, 93
Leverage ratios, 54, 58
Liquidity ratios, 54

M

Manufacturer's suggested retail price (MSRP), 42
Markup, 24, 28–31
Mathematical operations, 80, 87, 94, 110

N

Negative correlation, 84, 94–97, 106
No correlation, 84, 93, 102–105
Nonparametric statistical tests, 20
Normal curve
  definition, 16
  formula, 15
  generalizability factor, 18, 19
  normal distribution, 15
  t-distribution, 20, 21

theorem/law, 16, 17
Null hypothesis, 79, 80, 85, 109

O

One-tailed tests, 79

P, Q

Parametric statistics, 18
Pearson correlation
    characteristics, 84
    confidence level, 86, 93
    definition, 83, 84
    descriptive analysis, 105, 106
    Excel, 89
    hypothesis, 85, 93
    interpretation of results, 92
    mathematical operations, 87, 89
        negative correlation, 94–97
        no correlation, 102–105
        positive correlation,
          98, 99, 101
    results, 90, 92, 105
    statistical formula, 87, 89
Pie charts, 65, 75, 76
Positive correlation, 84, 92, 98–101
Probability
    general business
        examples, 26–28
    percentage, 24
    percent problems, 25
    problems, 23

properties, 24, 25
real-life percent, 31–33
real-life percent, profit
    margin, 33–35
real-life probability/percent
    examples, 28–30

R

Regression analysis, 91, 136
Retail Math
    basic metrics, 38, 41–43, 45, 46
    financial history, 37
    financial statement, 39
    growth metrics, 51, 52
    inventory/stock metrics, 47–50
    KPI, 38
Return on ad spending (ROAS), 131
Return on assets (ROA), 60, 61
Return on investment (ROI),
    131, 144

S

Salespeople, 144, 145
Shoppers, 68, 69, 72, 110, 118
Standard normal
    distribution, 18, 107
Statistical analysis, 72, 77,
    117, 145
Statistical procedure, 18, 78, 81
Statistical test, 21, 72, 78, 80
Student's t-test, 107

T, U

T-distribution, 8, 20, 107, 108
Two-tailed test, 20, 79, 85, 86,
    109, 120

V, W, X, Y, Z

Vertical bar
    chart, 73–74
Vertical charts, 71, 74

Printed in the United States
by Baker & Taylor Publisher Services

Printed in the United States
by Baker & Taylor Publisher Services